MODERN CHEMISTRY

Printed in Great Britain at the University Press, Cambridge
(Brooke Crutchley, University Printer)
and published by the Cambridge University Press
(Cambridge, and Bentley House, London)
Agents for U.S.A., Canada, and India: Macmillan

First edition 1946
Reprinted   1948

MODERN CHEMISTRY

Some Sketches of its Historical Development

BY

A. J. BERRY, M.A.

*Fellow of Downing College, and University Lecturer
in Chemistry, Cambridge*

CAMBRIDGE
AT THE UNIVERSITY PRESS
1948

The history of discovery may be held perhaps to supply the strongest reason for estimating effort towards clearness of thought as of not less importance in its own sphere than exploration of phenomena....

It is recognized of course that every attempt at improvement in scientific exposition must have a limited range, and that the chief critical interest will soon be transferred from what can be explained by any new formulation to what it has not shown itself competent to include.

Sir Joseph Larmor, Aether and Matter

La vraie méthode scientifique est très différente de celle que se représentaient les philosophes encore jusqu'à la fin du XIXe siècle et que l'on enseigne encore souvent actuellement. Le savant se sert de la logique des sentiments et de l'imagination tout autant que de la logique de son entendement. Entre la description exacte et systématique des faits et l'œuvre d'un savant il y a la même différence qu'entre la photographie d'un paysage et un tableau du même paysage fait par un artiste.

Victor Henri, Matière et Énergie

CONTENTS

ments according to their atomic numbers—discovery of element 72 predicted and subsequently realized as hafnium by Hevesy and Coster—elements 43 and 75—the last-named discovered as rhenium by Noddack—atomic numbers represent not only the ordinal numbers of the elements in the periodic table but also the number of the nuclear charge—verification of this by the scattering of α-particles by metals—the lanthanide contraction—J. J. Thomson's method of positive ray analysis—isotopes of neon—improvements in the apparatus introduced by Aston—accurate determination of atomic weights by the mass spectrograph—comparison with 'chemical' atomic weights—discovery of deuterium—partial separation of the isotopes of mercury by Brönsted and Hevesy—complete characterization of an element requires both atomic weight and atomic number—development of the modern theory of atomic structure—theories of valency of Kossel and of Lewis—extension of the latter by Langmuir—recognition of a third type of valency by Lowry and by Sidgwick—value of the Bohr atom as a basis for electronic theories of valency—theory of co-ordination compounds according to Sidgwick—resonance and its distinction from tautomerism.

PREFACE

It is gratifying to observe that interest in the historical development of science still continues to command attention, and indeed in some directions there is evidence of increased recognition of the importance of its study. The general history of chemistry has been written at various times, and there are several modern works which deal with the subject as a whole. Valuable as such books undoubtedly are for obtaining some sort of conspectus of the science from its early beginnings, the more recent developments in chemistry are of necessity relegated to the later chapters, and the treatment of them is therefore correspondingly curtailed.

The present volume is in no sense to be regarded as a substitute for any of these excellent works. Its purpose is rather to focus attention on the development of some of the newer branches of chemical science. As the result of supervising the studies of undergraduates for many years, the writer has had ample opportunities for seeing how seriously their understanding of important developments in modern chemistry suffers from the lack of any historical perspective of the evolution of the science, particularly in those branches in which progress has been most rapid since the close of the nineteenth century. It is certainly true that they are familiar with the names of many of the distinguished chemists of the past; but few of them seem to realize that the achievements of these 'old masters' with their limited resources and simple apparatus were little short of marvellous, and fewer still are in a position to understand that the study of the historical evolution of scientific theories is essential to a proper appreciation of their significance and import.

Accordingly an attempt has been made in the pages which follow to consider the development in historical perspective of certain branches of the science in separate chapters—each being nearly self-contained and largely independent of the others—and treated in a manner so as to be acceptable to anyone who is endowed with a moderate stock of chemical knowledge. To have attempted the writing of a more formal history of modern chemistry at the present time would have been a somewhat unsatisfactory task—it has been considered better to abandon some branches of the science altogether, rather than to attempt any sort of ill-digested completeness. In making a choice of subjects, the writer has been guided to some extent by his teaching experience. Thus organic chemistry, on account of its extensive and well-systematized character, would not have lent itself to satisfactory treatment in a work of this kind—the little organic chemistry which it contains, e.g. in the chapter on Stereochemistry, is really secondary to the main purpose of the book. In the chapter on Classical Atomic Theory the emphasis is on developments subsequent to Cannizzaro's generalization, since the confused state of chemical theory in the fifty years previous to that time has been adequately dealt with in other historical works on chemistry. The division of any science into departments must necessarily be to some extent arbitrary, but it is hoped that division according to branches of study rather than according to periods of time will be found more interesting, and certainly more convenient, from the point of view of the student. It is also hoped that the few references which are given will enable the reader to obtain direct or indirect access to original sources.

The author desires to express his sincere sense of gratitude to those friends who have most kindly assisted him in the production of this book. Dr E. A. Moelwyn-Hughes and Dr F. Wild have read nearly the whole of the typescript, and Dr Muriel Tomlinson, Fellow of Girton College, has read the proofs. The valuable criticisms and suggestions which they have made in the course of the work have been most welcome. The author would also extend his grateful thanks to the Staff of the University Press for their unfailing helpfulness, particularly in the difficult times in which this book has been produced.

A. J. B.

Cambridge
Michaelmas Term, 1944

Chapter I

CLASSICAL ATOMIC THEORY

It is difficult, indeed almost impossible, to assign precise dates to mark the beginnings of the various branches of chemical science, but it is fortunately less difficult to ascertain dates at which attention has been directed towards and interest aroused in some particular developments. Thus the atomic theory, associated with the name of Dalton as regards its bearing upon chemistry, is usually dated from the year 1808, but ideas regarding an atomic or discontinuous structure of matter were certainly prevalent in very much earlier times. Five years before this time, Berthollet had published his celebrated *Essai de Statique Chimique* which contained the germs of ideas destined to become of great importance half a century later in connexion with the beginnings of chemical dynamics, but which nevertheless contained serious errors regarding the constancy of composition of compounds. It was remarked by Lothar Meyer in the Introduction to the first edition of his *Modern Theories of Chemistry* (1864) that 'Dalton's speculations led him to a hypothesis which gave a surprisingly clear explanation of the doctrine of the fixity of composition of chemical compounds, a doctrine which was at that time disputed by Berthollet. This was the atomic hypothesis, which has since become the foundation of chemical science. The theory evolved from this hypothesis gave to chemistry quite a new form peculiarly its own.... It cannot be denied that, by the acceptance of the atomic theory, chemistry became more and more estranged from the nearly related science of physics....At the beginning of the present century, when chemistry guided by the atomic theory began its new and brilliant development, physics was not in a position to follow. Although the speculations of physicists were apparently based upon the existence of the smallest particles of matter, and although such expressions as molecules, pores, and interstices, were used by them, yet their calculations were in reality based upon the doctrine of the homogeneity and continuity of matter, and not upon the existence of single atoms, without the acceptance of which chemical teaching could no longer exist.'

Although general agreement will be given to Lothar Meyer's view that the development of chemistry around the atomic theory deflected the attention of chemists from the study of physics, it may be doubted if his remark regarding the contemporary state of physics was not an over-statement. It is true that developments resulting from the study of thermodynamics were not such as were likely to interest chemists in the early part of the nineteenth century, but the kinetic theory of gases, the modern development of which began with Clausius in 1857 and Clerk Maxwell in 1859, might have been expected to have exerted considerable influence upon the development of chemistry. As will be seen in what follows, the more recent developments in chemical science have been profoundly influenced both by the kinetic theory and by thermodynamics. Throughout the nineteenth century theoretical speculations had much less attraction for chemists than they presented to physicists, probably because of the success of the atomic theory, more especially since 1858 when Cannizzaro insisted upon the acceptance of Avogadro's hypothesis. Helmholtz in his Faraday Lecture in 1881 remarked that 'we have not yet any theory sufficiently developed which can explain all the facts of chemistry as simply and as consistently as the atomic theory'.

In 1808 Dalton published Part I of Volume I of his *New System of Chemical Philosophy*, and two years later he published Part II of the same work. In this treatise Dalton gave his atomic theory to the world. The year 1808 should also be remembered for the publication of Gay-Lussac's celebrated law of the combination of gases by volume—a purely experimental generalization. Three years later, in 1811, Avogadro published his hypothesis, the purpose of which was to provide an explanation of Gay-Lussac's law. The origin of Dalton's atomic theory has been discussed at various times since it was first formulated. Until the investigations of Roscoe and Harden and also of Debus in 1896, it was widely believed that the theory published in the *New System* was an attempt to explain chemical combination in constant and in multiple proportions, but such a view has been shown to be definitely erroneous. In 1896 Roscoe and Harden published their *New View of the Origin of Dalton's Atomic Theory*, which they had written as the result of a careful study of Dalton's own manuscript notes; and they concluded that the theory did not rest upon an experimental basis as had been generally supposed, but arose as the result of Dalton's attachment to Newton's ideas regarding an atomic constitution of matter as expressed in the *Principia*. Newton's advocacy of the corpuscular theory of light in the *Opticks* shows how much committed he was to atomic conceptions. Somewhat similar views regarding the origin of the theory were expressed by Debus between 1896 and 1899, but Debus also regarded some experiments which Dalton had carried out on the oxides of nitrogen as having had some bearing on the development of his theoretical ideas.

Gay-Lussac's law of volumes received a satisfactory explanation three years afterwards by the publication in 1811 of Avogadro's hypothesis, and it is generally accepted that the theorem originated in this way. A very different view was, however, expressed by Debus on this subject. Debus considered that the theorem universally associated with the name of Avogadro should be attributed to Dalton, and even asserted that the principle had been deduced by Dalton as early as 1801. Dalton had been making experiments on the combination of gases by volume on similar lines to those of Gay-Lussac, but his work was much inferior to that of his French contemporary as regards accuracy. Nevertheless, Dalton rejected Gay-Lussac's law and also an attempt to explain the results theoretically as early as 1805. In any case it is very difficult to understand how Dalton could uphold Avogadro's theorem and at the same time reject Gay-Lussac's law. This claim by Debus on Dalton's behalf has not been made by others who have gone into the history of the atomic theory. In particular, Meldrum, in a monograph entitled *Avogadro and Dalton* published in 1904, has definitely rejected it.

Of the four systems of chemistry which came into existence in the nineteenth century, the first was that of Berzelius. A preliminary system was brought out by him in 1810 and an improved one in 1826. In establishing his table of atomic weights,

Berzelius showed himself to be a master of analytical chemistry, but his selection of general principles for fixing the atomic weights was not altogether logical. The chief principles upon which he relied included Dulong and Petit's law formulated in 1819, and Mitscherlich's law of isomorphism enunciated in 1821. Berzelius also made some use of Gay-Lussac's law of volumes, but in such a way as to indicate that equal volumes of gaseous elements contained the same number of atoms, not in Avogadro's sense. In any case it should be noted that although there is a fairly close resemblance between the modern formulae for some simple compounds and the formulae of Berzelius, the similarity was largely fortuitous. A different system of atomic weights, which would be more correctly termed combining or equivalent weights, was published by Gmelin in his *Handbuch der Chemie* in 1843. This theory was fundamentally a gravimetric one. Gmelin seems to have realized some of the difficulties to be encountered in attempting to reconcile the atomic theory with volume relations, and he therefore restricted his method for arriving at formulae and atomic weights chiefly to the results of analysis and to considerations of isomorphism. It would appear that in a sense Gmelin had returned to the conceptions of equivalents which Wollaston had suggested as early as the year 1814.

The invention by Dumas in 1826 of a useful method for determining the vapour density of volatile substances had considerable influence on the subsequent progress of chemistry in several directions. By accepting Avogadro's rule it provided the means of determining the molecular weights of volatile substances in the vaporous condition. At this time increased interest was being taken in organic compounds, and thus arose a system of chemistry which took a more serious account of volume relationships. This was the system published by Gerhardt in his *Traité de Chimie organique*, and it achieved a high degree of success with carbon compounds, but with inorganic substances it was somewhat of a failure. This important work appeared in four volumes between 1853 and 1856, and in Volume IV of the *Traité* Gerhardt developed his theory of types—an idea which had previously been adopted in more than one way by others, such as Dumas and Williamson—according to which the formulae of substances could be referred to four typical compounds, namely, water, hydrochloric acid, hydrogen, and ammonia. It is interesting to note that Gerhardt took account of Wurtz's discovery in 1849 of substituted ammonias, afterwards elaborated in greater detail by Hofmann, and understood the analogy between ethylamine and ammonia. Laurent, whose name has always been closely associated with that of Gerhardt, embodied his theoretical ideas in a volume entitled *Méthode de Chimie* and published after his death. An English translation of his work was issued by Odling in 1855. With both Gerhardt and Laurent the molecule, much more than the atom, was the fundamental unit in chemistry, and it is certain that the revival of interest in Avogadro's theorem was very largely brought about by their activities.

The confused state of chemistry up to the middle of the nineteenth century is perhaps more easily imagined than described. Different formulae were used by different chemists for simple compounds, and there was no common system of atomic weights. The final step towards the establishment of a uniform and consistent system of chemistry was taken by Cannizzaro in 1858. This he did by insisting upon the *universal* applicability of Avogadro's hypothesis as a basis for the determination of molecular weights, and from this he was able to derive formulae and atomic weights free from the inconsistencies which abounded in the other systems of chemistry.

It is interesting to observe that the older writers on chemistry, almost without exception, used to refer to Avogadro's *hypothesis*, whereas in more modern times there has been a marked tendency to designate the principle as Avogadro's *law*. On historical grounds there is not the slightest doubt that the term hypothesis is the correct one. But as every subsequent step in the progress of chemical science has directly or indirectly provided evidence of the correctness of the principle, the use of the term law has become well established. Thus Mendeleeff in his *Principles of Chemistry* (third English edition, 1905) summarized the position by stating that 'the hypothesis must be considered as verified and the law of Avogadro-Gerhardt must be spoken of as fundamental'. The association of Gerhardt's name with Avogadro's rule was stressed by Mendeleeff. It might be added that measurements of the constant known as Avogadro's number, namely, the number of molecules in a gramme-molecule, by numerous widely different methods all lead to figures showing a most remarkably close agreement, and consequently the use of the term law in connexion with Avogadro's theorem is now amply justified.

The development of the fundamental ideas on valency can be traced as far back as the year 1819 when Berzelius introduced his electrochemical theory of chemical union, also known as the dualistic theory of combination. Closely connected with this subject is the gradual evolution of the idea of groups or radicals, namely, aggregates of two or more atoms which can take part in a number of reactions without losing their integrity or individuality as a whole. The first example of this kind appears to have been the radical cyanogen discovered by Gay-Lussac in 1815, who recognized its analogy to chlorine. In 1824 Liebig and Wöhler demonstrated the identity of cyanic and fulminic acids as regards composition, and thus brought forward one of the very early examples of isomerism, a term which appears to have been first used by Berzelius to denote identity between substances as regards composition combined with differences in properties and reactions. Modern organic chemistry is usually said to date from the year 1828 in which Wöhler effected the transformation of ammonium cyanate into urea, and this has been frequently quoted as the first example of the synthesis of a vital product. Very shortly before this time, actually in 1826, Hennell had accomplished the synthesis of ethyl alcohol from ethylene by absorption in sulphuric acid and subsequent hydrolysis of the product. Hennell must therefore be claimed equally with Wöhler as one of the pioneers in the synthesis of natural products. The immediate developments in organic chemistry after that time were very much concerned with the study of radicals, such as that of benzoic acid, discovered by Liebig and Wöhler in 1832, and that of the cacodyl compounds by Bunsen about 1841. The evolution of the theories of types and the development of the doctrine of valency arose as a consequence of these discoveries. Contributions of outstanding importance to the theory of valency were made by Williamson's researches on etherification between 1850 and 1856 and by those of Frankland on organo-metallic compounds about 1852. In 1858 Couper introduced formulae with dotted lines to represent the valency bonds, and in the same year Kekulé laid the foundations of structural chemistry on the basis of constant valency.

An important question regarding the constancy or variable nature of valency was widely discussed during the middle years of the nineteenth century, and opposite views on this question were held by Kekulé and Frankland. Kekulé was firmly of opinion that the valency of an element was fixed and unalterable, and in developing this view, the success of which as regards organic chemistry is disputed by no one, he drew a distinction between 'atomic' and 'molecular' compounds. Thus to preserve the elements nitrogen and phosphorus as universally tervalent, Kekulé wrote the formulae of ammonium chloride and of phosphorus pentachloride as NH_3, HCl and as PCl_3, Cl_2 respectively. It may be noted that this way of expressing formulae is still to be seen in the formulae of compounds such as aniline hydrochloride which is usually written as $C_6H_5NH_2$, HCl. Kekulé's formulae for compounds such as ammonium chloride and phosphorus pentachloride was a useful method of giving expression to the readiness with which such compounds undergo thermal dissociation, but at the same time it resulted in raising the very difficult question as to where molecular compounds end and atomic compounds begin. It may fairly be said that the doctrine of constant valency as laid down by Kekulé had a great deal to do with the rapid rise and development of organic chemistry, but at the same time the views of Frankland regarding variable valency were found to be correct as regards many of the elements, and thus gradually secured universal recognition.

Apart from an important generalization noted by Mendeleeff regarding valency at the time when he drew up the periodic table, relatively few attempts were made to correlate theories of chemical union with ideas on valency until the later years of the nineteenth century. For the most part chemists were content to recognize that in some elements, such as carbon, hydrogen, and oxygen, the valency was apparently constant, whereas in others, such as nitrogen, phosphorus and sulphur, the valency was variable. Gradually, however, it was found difficult to maintain even such elements as iodine and oxygen as having a constant valency. Thus it was found necessary by Victor Meyer in 1894 to assume that iodine was tervalent in compounds such as diphenyliodonium hydroxide, $(C_6H_5)_2IOH$, in which the halogen displayed the very unusual property of acting as a base-forming element. The possibility of oxygen acting as a quadrivalent element was first recognized in the addition compound of dimethyl ether and hydrogen chloride, which was investigated by Friedel as long ago as 1875, but much more definite evidence was forthcoming from Collie and Tickle's study of the salts of dimethylpyrone in 1899. The hydrochloride of this compound was formulated by them as

$$CO \begin{matrix} CH{=}C(CH_3) \\ \\ CH{=}C(CH_3) \end{matrix} O \begin{matrix} H \\ \\ Cl \end{matrix},$$

and may be regarded as an example of the so-called oxonium salts as understood by Baeyer and Villiger, who discussed the basic function of oxygen in certain types of organic compounds in 1901. They pointed out that the types of structure which raise or diminish the basic properties of oxygen are identical with those which produce effects of this kind in compounds which contain nitrogen, and the term *oxonium* was introduced to give expression to this similarity with ammonium salts. Thus the number of elements of unvarying valency gradually diminished, and since 1892 a series of memoirs was published by Nef in which he attacked the theory of the universal quadrivalency of carbon. It is by no means surprising that Nef's views were extremely unpopular with the majority of organic chemists, most of whom were strong adherents of the Kekulé school of thought. Nevertheless, Nef's views regarding the bivalency of carbon in compounds such as carbon monoxide and isonitriles were of great interest and have received attention from a later generation of chemists.

Most valuable ideas, originally of a purely classificatory character, were put forward by Werner from 1891 onwards regarding valency in certain complicated types of inorganic compounds. His object was to introduce some sort of order and methodology into the formulae of compounds with water of crystallization and ammine salts. In the light of the subsequent developments in inorganic chemistry, it might almost be said that Werner's ideas have been of scarcely less theoretical importance than those of Kekulé and van't Hoff. Between 1899 and 1904 problems of valency were approached from an electro-chemical standpoint by Abegg, whose views might be described as an attempt to correlate the ideas of Mendeleeff with the earlier dualistic electrochemical conceptions of Berzelius.

Considerable interest is associated with the history of Ber-zelius's dualistic theory of chemical combination, more par-ticularly as regards its downfall. Its origin is doubtless to be traced in the electrochemical researches at the beginning of the nineteenth century with which the name of Davy was so bril-liantly associated. According to Berzelius, salts were to be regarded as the products which result from the union of a basic oxide with an acidic oxide, the basic oxide being the positive and the acidic oxide the negative constituent of the salt. For a number of years this theory enjoyed a considerable measure of support, and this is by no means surprising when it is remem-bered that at that time there was a fairly considerable edifice of inorganic chemistry, whereas organic chemistry had scarcely begun to raise its head. At least two major causes must be assigned to account for the decline and fall of the dualistic theory. First, the recognition of the elementary nature of chlorine by Davy, whereas Berzelius had regarded the gas as an oxidized hydrochloric acid, and indeed had called it 'oxymuriatic acid'. It was thus possible to have chlorine as the electronegative constituent of a salt instead of an acidic oxide. The dualistic theory was thus only applicable to the salts of oxy-acids. Secondly, there was the gradual increase of interest in the de-velopment of organic chemistry, particularly the discovery by Dumas in 1834 that hydrogen can be substituted in organic compounds by chlorine, atom by atom, without thereby pro-ducing very profound changes in the properties of the resulting products. The electrochemical theory of combination thus received a severe setback by the observation that a strongly electronegative element like chlorine could replace an admit-tedly electropositive one such as hydrogen in the molecule of a compound without a corresponding change in the character of the substance thus produced. A third cause may be suggested as having had some influence upon the decline of the dualistic theory. The great Swedish chemist had deservedly won a high reputation for accuracy in inorganic analysis, and his atomic weights had been accepted almost without question. In 1841, however, a serious flaw was discovered by Dumas and Stas in the value assigned by Berzelius for the atomic weight of carbon in 1839. They found that his value was about 2 per cent too high, and obtained a much more accurate figure. The discovery of this error in the atomic weight of carbon had the effect not only of depressing the high reputation which the atomic weights of Berzelius very rightly enjoyed, but it exercised an indirect influence upon the downfall of the dualistic theory. Although this theory as originally formulated has long ceased to attract attention, except perhaps as a useful means of studying the composition of silicates, modern electrical theories of chemical union have undoubtedly sprung from the roots of Berzelius's electrochemical conceptions. It should not be forgotten that the value of any theory depends less upon its truth than upon its usefulness.

At various times doubts were cast by some chemists regarding the universal applicability of Avogadro's theorem. These doubts originated in connexion with the so-called abnormal vapour densities which were exhibited by substances such as ammonium chloride, ammonium carbamate, phosphonium bromide and phosphorus pentachloride, the molecular weights of which as determined in the vaporous condition would have led to illogical formulae. The correct explanation of these supposedly abnormal vapour densities was given in 1857–8 independently by Canniz-zaro, Kopp, and Kekulé. Nevertheless, it is curious to reflect that some of the French experimentalists on thermal dissocia-tion, the chief of whom was Sainte Claire Deville, who had thus brought forward a considerable volume of evidence in favour of the correctness of Avogadro's hypothesis, rejected the pheno-mena of abnormal vapour densities being explained on the basis of the dissociation of the molecules. The correctness of that explanation, particularly as regards phosphorus pentachloride, was demonstrated by Wurtz in 1869.

The problems of molecular weights in the vaporous condition and of thermal dissociation are closely interwoven, and occupied much attention during the nineteenth century. On the experimental side the method of Dumas was supplemented about 1868 by a method due to Hofmann, which was an adaptation of an earlier one due to Gay-Lussac, and in 1877 Victor Meyer devised his celebrated displacement method. As a means of determining molecular weights of an approximate degree of accuracy Victor Meyer's method has proved extremely useful, but it is less satisfactory for the study of thermal dissociation than the older method of Dumas, because the products of dissociation become diluted with air, their partial pressures are thereby reduced, and consequently the degree of dissociation is artificially increased. On the theoretical side much attention was given to the problem of the molecular formulae of substances, the vapour density of which was found to be constant over a certain range of temperature and then gradually to fall to a lower value, often to one-half of the previous figure. Thus Deville and Troost in 1860, who used Dumas's method, and Nilson and Pettersson in 1887, who used the method of Victor Meyer, found that the vapour density of aluminium chloride at temperatures below 500° C. corresponded to a molecular formula of Al_2Cl_6, whereas the value at temperatures above 1000° C. was in agreement with a formula of $AlCl_3$. Numerous other examples of the same kind might be quoted, and the results gave rise to much discussion regarding the valency of the metal in such compounds. Eventually it was realized that there was no question of one formula being right and the other wrong. Both formulae are correct within the particular range of experimental conditions in which they are obtained.

The use of structural or constitutional formulae to give expression to the properties and reactions of compounds and to explain the existence of isomerides arose as the result of the work of a number of chemists, particularly Frankland, Couper, Kekulé, and Kolbe. This subject was closely interwoven with the rapid development of organic chemistry on the experimental side, but much of the confusion which abounded was due to the use of Gmelin's atomic weights instead of those of Cannizzaro. Until the final acceptance of Cannizzaro's teaching there was no consistency as regards the structural formulae to be assigned to relatively simple molecules. Thus the hydrocarbon ethane was regarded by Frankland in 1849 as ethyl hydride, because it could be obtained by reducing ethyl iodide with zinc and water. The hydrocarbon which Frankland obtained by the action of zinc upon dry ethyl iodide was termed ethyl by him, but it is really butane. The identity of ethyl hydride with the substance known as dimethyl, which Wurtz prepared by the action of sodium upon methyl iodide, seems to have been foreshadowed by Brodie and was finally established by Schorlemmer in 1865. Neglect of modern atomic weights did not, however, prevent Kolbe from recognizing isomerism among alcohols. About 1860 he indicated that the three types of isomeric alcohols could all be regarded as derivatives of carbinol, and thus introduced an extremely convenient nomenclature for characterizing these classes of compounds. In this system the primary, secondary, and tertiary alcohols were regarded as carbinols in which one, two, and three hydrogen atoms in the molecule respectively were replaced by organic groups.

Ever since chemists made use of formulae to represent the molecules of substances and their reactions, opinions have differed considerably regarding the nature and quality of the information imparted by the formulae. Is there but one 'correct' formula for a substance, and are other formulae erroneous, or at least mere approximations to the truth? The anarchy which prevailed in the early years of the nineteenth century when there were four 'systems' of chemistry, happily ended as the result of Cannizzaro's teaching, would seem to indicate that once certain general principles are accepted, consistent formulae would follow as a logical consequence. In the *Traité de Chimie organique*, Tome IV, a contrary view was expressed by Gerhardt. Thus he stated (p. 580): 'Le principe qu'*un seul et même corps peut avoir deux ou plusieurs formules rationnelles* sera sans doute contesté par les chimistes qui prétendent représenter par les formules chimiques la constitution absolue des molécules; il ne saurait, au contraire, être nié par ceux, comme moi, ne voient dans les formules qu'un manière de concréter certains rapports de composition et de décomposition. Je dis plus: en immobilisant en quelque sorte un corps dans une seule formule, on se cache souvent à soi-même des relations chimiques dont une autre formule donne immédiatement la perception....' Nearly thirty years later, in 1885, the same question was raised by Conrad Laar in a paper entitled 'Über die Mögligkeit mehrerer Strukturformeln für dieselbe chemische Verbindung', in which he drew attention to several examples of what is known as tautomerism, including the classical example of ethyl acetoacetate discovered by Geuther in 1863, who regarded the substance as β-hydroxycrotonic ester, whereas Frankland considered it to be a ketonic compound.

Laar's explanation of the behaviour of compounds such as acetoacetic ester consisted in the idea of a mobile hydrogen atom oscillating between two different atoms, usually between a carbon atom and an oxygen atom, within the same molecule, so as to give expression to the two-fold chemical behaviour of the compound, and the term tautomerism was understood by him in that sense. At an earlier date, actually in 1877, Butleroff, as a result of his investigation of the action of sulphuric acid upon tertiary butyl alcohol, recognized the existence of two isomeric *iso*dibutylenes

$$(CH_3)_3C{-}CH{=}C{\begin{array}{c} CH_3 \\ CH_3 \end{array}} \quad \text{and} \quad (CH_3)_3C{-}CH_2{-}C{\begin{array}{c} CH_3 \\ CH_2 \end{array}}$$

and explained their formation by isomeric changes taking place reversibly through the intermediate formation of the alcohol. Butleroff thus recognized the possibility of reversible isomeric change, and his views were certainly more in agreement with those which have prevailed in more modern times than the assumption of an oscillating hydrogen atom as understood by Laar. And in 1872, Kekulé, having recognized the difficulties which would arise in consequence of the non-existence of isomerism in *ortho* disubstituted benzenoid compounds, introduced his celebrated dynamic formula for benzene in which he regarded the double bonds in the molecule as being in a state of continual oscillation. This conception has also been described as intra-annular tautomerism in order to distinguish it from tautomerism connected with the movement of a hydrogen atom and thus resulting in intramolecular change.

In 1896 Claisen was able to bring forward direct experimental evidence in favour of the correctness of the views of Butleroff based upon the principle of reversible isomeric change as opposed to the ideas of Laar. Thus he was able to prepare certain diketones and ketonic esters in the isomeric enol and keto forms and isolate them in crystalline condition. One of the first examples of this kind to be obtained were the two forms of acetyldibenzoylmethane,

$$CH_3.CO.CH(CO.C_6H_5)_2 \quad \text{and} \quad CH_3.C(OH):C(CO.C_6H_5)_2,$$

each of which could be converted into the other. And in the same year Hantzsch and Schultze succeeded in isolating phenyl-nitromethane in two isomeric forms which were mutually convertible. One of these is the true nitroparaffin $C_6H_5.CH_2.NO_2$ and the other is the acidic isomer $C_6H_5.CH:NO.OH$, which has an ionizable hydrogen atom replaceable by metals. This discovery was connected with the early beginnings of the theory of pseudo-acids and pseudo-bases, which has assumed much importance in more recent years. Gradually a considerable number of compounds were recognized as readily capable of undergoing conversion reversibly into isomeric forms; and in many instances, but by no means in all, the corresponding isomerides were eventually discovered. Wide variations were found as regards the rapidity of transformation in different cases, and thus arose the array of terminology used by different writers at various times. The whole subject was first considered comprehensively by Lowry in 1899, who substituted the term dynamic isomerism for terms such as tautomerism and desmotropy, so as to include all phenomena of this kind whether the isomerides were known or were still undiscovered. Thus it was generally recognized that ethyl acetoacetate was not a single substance but an equilibrium mixture long before a complete experimental demonstration was forthcoming in 1911, when the two isomerides were isolated by Knorr. The term tautomerism has, however, survived, not in Laar's sense, but as synonymous with *rapid* isomeric change.

Apart from the necessity of obtaining values for the atomic weights of a degree of accuracy sufficient for purely practical purposes, a stimulus to research on atomic weights was provided by a hypothesis due to Prout (1815), that the values were all integral numbers. Of the numerous investigations of this kind, particular interest is attached to a research by Penny in 1839, who determined 'several equivalent numbers' by evaporating a known weight of a salt with successive quantities of a volatile acid and estimating the weight of the resulting product. Thus chlorates were converted into chlorides, chlorides into nitrates, and nitrates reciprocally into chlorides. The values published by Penny were oxygen 8, chlorine 35·45, nitrogen 14·02, potassium 39·08, and silver 107·97, and he remarked significantly that 'the favourite hypothesis of all equivalents being simple multiples of hydrogen is no longer tenable'. Mention must also be made of experiments of a high degree of accuracy carried out by Marignac in 1842 and 1843 on the ratio of silver to oxygen. The important researches of Stas from 1860 onwards, were very much concerned with the preparation of highly pure silver. Indeed, if oxygen is to be called the primary standard in atomic weight research, it is equally correct to designate silver as the secondary standard, because equivalent weight determinations can be effected with greater accuracy between a halogen and silver than in other ways. Although it is true that Stas's experiments were carried out with the greatest care, the accuracy of some of his values was surpassed by those of others; for example, the value which for long was assigned by him to nitrogen, viz. 14·04, is less accurate than Penny's figure.

A considerable part of Stas's atomic-weight work was concerned with the determination of the ratio of silver to oxygen, and this he obtained in more than one way. Thus by converting potassium chlorate into the chloride, and this latter salt into silver chloride, and by converting silver directly into the chloride he was able to determine this ratio, and eventually to derive the several atomic weights of the elements in these compounds. It should be noted that this important ratio was determined indirectly. All attempts to determine the ratio of silver to oxygen directly were found to be unsatisfactory because of the great difficulty of preparing silver oxide in a condition of purity—an operation which was finally accomplished by Riley and Baker as late as 1926. Their value supported the results of previous determinations by the indirect methods.

It would appear that although Stas achieved a very high degree of success in the preparation of pure substances, the values of his atomic weights were somewhat exaggerated. The causes of this were largely discovered as the result of the revision of many of the atomic weights by Richards and his collaborators. Stas undoubtedly worked with too large quantities of material and in too concentrated solutions. Errors thus arose in consequence of the adsorption of substances on the precipitates. Another source of error arose in connexion with Stas's practice of fusing his silver under an oxidizing flux and thereby causing it to occlude oxygen. Richards pointed out that silver required for atomic-weight work must be fused in a hydrogen atmosphere or *in vacuo*.

The atomic weight of nitrogen has been determined in many different ways, and is an interesting example of a value, long regarded as beyond question, being doubted and finally improved as the results of the refinements introduced into this kind of work by Richards. One method which has been used at widely separated dates, namely, the conversion of silver into silver nitrate, has given the following values on the basis of 107·92 for the atomic weight for silver: Penny in 1839 found nitrogen to have an atomic weight of 13·99, Stas in 1860 obtained a value of 14·03, about 1865 his value was 14·05, and in 1907 Richards and Forbes found the atomic weight to be 14·03.

These figures all rest upon the accuracy of the value of 107·92 for the atomic weight of silver. But determinations of the atomic weight by the method of measurement of the limiting density of the gas as carried out by Guye, Lord Rayleigh, and others up to the year 1905 (Chapter VI) had given values for the atomic weight of nitrogen of 14·009. In 1906 Whytlaw Gray made a critical analysis of Stas's nitrogen ratios with the object of reconciling the value of 14·04, which was the then accepted value for the atomic weight of the element as determined by 'chemical' methods with that found by the limiting density method. He concluded that Stas's nitrogen ratios were not wholly flawless, and in particular that his earlier values for the atomic weight were the more accurate. The discrepancy was finally cleared up in 1909 by Richards in collaboration with Köthner and Tiede who redetermined the combining ratio of ammonium chloride to silver chloride—a ratio previously used by other workers on atomic weights—and found that the correct value for the atomic weight of silver was 107·88, the corresponding value for nitrogen being 14·008.

Attempts to derive relations between the numerical values of the atomic weights of similar elements and their properties were made during the early part of the nineteenth century. Thus Döbereiner as early as 1829, and Dumas some thirty years later, drew attention to elements in groups of three, the atomic weight of the middle element being approximately the arithmetic mean of that of the two extreme elements. Thus bromine with an atomic weight approximating to 80 has a value which is in the neighbourhood of the mean of those of chlorine (35·5) and iodine (127). Similar considerations apply to the atomic weight of strontium (87·5), which is not far removed from the value of the arithmetic mean of that of calcium (40) and that of barium (137). Relationships of this kind were known as the law of triads. It may be noted that such a rule does not go very far, and further it is not difficult to find elements having atomic weights which are roughly the arithmetic mean of those of others without any chemical similarity whatever. Nevertheless, there is no doubt that these earlier speculations were not without influence upon the subsequent development of the periodic system of classification.

In 1862 de Chancourtois published his celebrated spiral of the elements in which he showed that if the values of the atomic weights were written along the generatrix of a vertical cylinder, having the circular base divided into 16 equal parts, and a helix traced on the cylinder at a constant angle of 45° to its axis, a relationship between the values of the atomic weights and the nature of the elements becomes apparent. The *characteristic numbers* (atomic weights) measured off on the spiral caused analogous elements to fall on the same vertical lines. This interesting idea, known as the telluric screw, was almost wholly ignored.

Between 1863 and 1866 a series of short papers was published in *The Chemical News* by Newlands on numerical relations between the atomic weights of the elements and their corresponding properties. He found that if the elements, omitting hydrogen, were arranged in the order of ascending atomic weights, 'the numbers of analogous elements generally differ either by seven or by some multiple of seven', like 'the extremities of one or more octaves in music'. This relation was termed the law of octaves by Newlands. It was read before the Chemical Society in 1866 and publication was refused.

Although it is almost universally admitted that germs of the periodic law are to be found in the works of de Chancourtois and of Newlands, it appears that the discoverer of that most important generalization, Mendeleeff, was unacquainted with their publications, but he was nevertheless influenced by the works of others, for example, Dumas. Mendeleeff enunciated the periodic law in 1869, and he stressed particularly that (1) the elements if arranged according to their atomic weights exhibit an evident *periodicity* of properties, (2) elements which are similar as regards their chemical properties have atomic weights which are either of nearly the same value or which increase regularly, (3) the arrangement of the elements or groups of elements in the order of their atomic weights corresponds with their so-called valencies, (4) the elements which are most widely distributed in nature have *small* atomic weights, and all the elements of small atomic weights are characterized by sharply defined properties, (5) the magnitude of the atomic weight determines the character of an element, (6) the discovery of many yet unknown elements may be expected, (7) the atomic weight of an element may sometimes be corrected by the aid of a knowledge of those of the adjacent elements, and (8) certain characteristic properties of the elements can be foretold from their atomic weights. Twenty years later, Mendeleeff delivered the Faraday Lecture before the Chemical Society and repeated these eight generalizations in almost the original words.

In 1870, Lothar Meyer independently put forward what was fundamentally the same idea, namely, that the properties of the elements and their compounds are related to the values of the atomic weights. His generalization is strikingly demonstrated by the well-known curve in which the atomic weights are plotted as abscissae against the atomic volumes of the elements as ordinates. The periodic character of the curve is immediately obvious. Lothar Meyer is often quoted as an independent discoverer of the periodic law. His contributions to the subject must, however, be regarded as inferior to those of Mendeleeff, as the

Early form of periodic table (after Mendeleeff)

Series	Group I — R^2O	Group II — RO	Group III — R^2O^3	Group IV RH^4 RO^2	Group V RH^3 R^2O^5	Group VI RH^2 RO^3	Group VII RH R^2O^7	Group VIII — RO^4
1	H=1							
2	Li=7	Be=9·4	B=11	C=12	N=14	O=16	F=19	
3	Na=23	Mg=24	Al=27·3	Si=28	P=31	S=32	Cl=35·5	
4	K=39	Ca=40	—=44	Ti=48	V=51	Cr=52	Mn=55	Fe=56, Co=59, Ni=59, Cu=63
5	(Cu=63)	Zn=65	—=68	—=72	As=75	Se=78	Br=80	
6	Rb=85	Sr=87	?Yt=88	Zr=90	Nb=94	Mo=96	—=100	Ru=104, Rh=104, Pd=106, Ag=108
7	(Ag=108)	Cd=112	In=113	Sn=118	Sb=122	Te=125	I=127	
8	Cs=133	Ba=137	?Di=138	?Ce=140	—	—	—	— — — —
9	(—)	—	—	—	—	—	—	
10	—	—	?Er=178	?La=180	Ta=182	W=184	—	Os=195, Ir=197, Pt=198, Au=199
11	(Au=199)	Hg=200	Tl=204	Pb=207	Bi=208	—	—	
12	—	—	—	Th=231	—	U=240	—	— — — —

great Russian chemist predicted that atomic weights which did not fit into his table would be found to be erroneous, and also that gaps in his table would eventually be filled by hitherto undiscovered elements. Lothar Meyer's whole attitude to the subject was much more restrained and cautious, and in particular he was opposed to altering atomic weights which could not be fitted into the system. It will be seen in what follows that the attitude taken up by Mendeleeff was fully justified; several elements of doubtful or incorrect atomic weight were subsequently incorporated into the table with their values corrected, and in three instances the properties of elements unknown at that time, predicted and termed *ekaboron, ekaaluminium,* and *ekasilicon,* were subsequently realized by the discovery of scandium by Nilson in 1879, of gallium by Lecoq de Boisbaudran in 1875, and of germanium by Winkler in 1886. A distinctive feature of Mendeleeff's periodic table is that every element occupies a certain position, which is determined by the group (indicated by Roman numerals) and series (Arabic numerals) where it occurs. Vacant spaces were left to be filled by elements the properties of which could be foretold by considering the properties of the elements around any particular space.

Considerable importance was attached by Mendeleeff to the periodicity of the valency of the elements according to his system of classification. He pointed out that the elements are capable of combining with a greater amount of oxygen the less the amount of hydrogen with which they can combine; and he went on to state that the sum of the equivalents of hydrogen and oxygen of what he termed the higher types is equal to eight. This principle was reproduced by Abegg some thirty years later, expressed in electrochemical language, in his theory of normal and contravalencies. A most interesting example of assigning valencies to elements, having regard to the properties of their oxides, is to be seen in Mendeleeff's views on the rare earths. The oxides of elements such as erbium, yttrium, cerium, and 'didymium' were considered by several chemists, by Bunsen for example, to be of the RO type like the alkaline earths on account of their strongly basic character. In 1871, however, Mendeleeff changed the formulae of these earths, including that of cerous oxide, to the R_2O_3 type so as to fit them into the table. For lanthana he gave the erroneous formula LaO_2. How often has the periodic table been reconstructed since Mendeleeff's time!

Although the rare earths have been a real difficulty in the periodic system until recently, their study has not been without considerable influence in the development of this classification. An interesting example is to be found in the events which led to the discovery of scandium. The beginnings of the study of the rare earths may be traced back to the discovery of the earth yttria by Gadolin in a Scandinavian mineral in 1794 and of the earth ceria in cerite by Berzelius and Hisinger in 1803. This earth was examined independently by Klaproth. Up to the middle of the nineteenth century numerous investigations on the separation of these earths into others, of which mention may be made of the work of Mosander (1839–43), which led to the discovery of lanthana and 'didymia', and that of Cleve (1873–4), which verified and extended the views of Mendeleeff regarding the valency of the elements. Cleve showed that all the oxides, including lanthanum oxide, to be of the R_2O_3 type, and the tervalency was further confirmed by Hillebrand and Norton in 1875, who determined the specific heats of the metals cerium, lanthanum, and didymium. In 1878 the chemistry of the rare earths entered upon a new phase by making use of the mineral samarskite, found in large quantities in North America, as a source for preparing them. It was, however, in gadolinite that Marignac in this year discovered a new earth which he called ytterbia, and assigned an atomic weight of 172·5 to the element ytterbium. In 1879 Nilson repeated Marignac's work and raised the atomic weight to 173. In the course of his investigation Nilson observed the separation of a less basic oxide, the metal having a much lower atomic weight, namely, 44, identical with the figure previously predicted by Mendeleeff for the element *ekaboron*.

The value of the periodic law in controlling or correcting doubtful atomic weights is now well known and may be illustrated with reference to beryllium. The equivalent weight of this element is 4·5. If beryllium is compared with magnesium and with aluminium the element and its compounds resemble the latter element rather than the former, and on that account it was considered by some chemists to be tervalent, but such irregularities were familiar to Mendeleeff. The bivalency of beryllium was established chiefly by the researches of Nilson and Pettersson (1880–4), who found that if the specific heat of the metal was determined at sufficiently high temperatures, the value of the atomic heat was consistent with the position assigned to beryllium in Group II. This was confirmed by determinations of the vapour density of the anhydrous chloride. Nilson and Pettersson employed Victor Meyer's method, which had been devised shortly before, but they found that great care was necessary to obtain satisfactory results, and were obliged to have recourse to the older method of Dumas for studying the transitory state in which the molecules were partially polymerized. From 730° C. upwards results consistent with the molecular formula $BeCl_2$ were obtained and the bivalency of the element definitely established.

For many years the atomic weight of tellurium was a constant source of difficulty. As regards the position of the element in the periodic table, there was no question that its chemical properties settled its position in Group VI along with sulphur and selenium. But values of the atomic weight determined up to the time of the formulation of the periodic law approximated to 128–129, and were thus definitely higher than that of iodine, which according to Stas (1865) was 126·8. Even at that time it was admitted that much greater weight should be placed upon the reliability of the iodine value than of that of tellurium, on account of difficulties in obtaining pure preparations. Mendeleeff was well aware of the discrepancy, and was convinced that tellurium would be found to contain an element of higher atomic weight, and he called this hypothetical element dwi-tellurium with an atomic weight approaching a value of 212. This problem was investigated by several chemists, notably by Brauner, who carried out a most elaborate series of experiments in purifying tellurium compounds between 1883 and 1889, and determined the atomic weight by more than one method. All his most reliable work gave values for the atomic weight higher than that of iodine, and actually ranging between 127·5 and 127·7. The problem was then attacked by Scott in 1902, who employed trimethyltellurium iodide and trimethyltellurium bromide for his determinations. By converting the iodine in the former compound gravimetrically into silver iodide he obtained a value of 127·6, and by titrating the bromine in the latter compound by Stas's silver method his value for the atomic weight was 127·7. Scott also redetermined the atomic weight of iodine and obtained a value of 126·92. A most thorough and exhaustive investigation was carried out by Baker and Bennett in 1907. These investigators obtained their material from several different parts of the world, and they employed various methods for purifying their tellurium preparations. One of their methods for determining the atomic weight was an interesting adaptation of a method which had been employed nearly a century before by Berzelius for determining the atomic weight of arsenic, and consisted in heating tellurium dioxide with excess of sulphur in such a way that only sulphur dioxide escaped from the apparatus. By correlating the loss of weight due to sulphur dioxide with the weights of tellurium dioxide used, a value of 127·6 was obtained. Other analytical methods employed by Baker and Bennett gave essentially the same result. It might have been supposed that the results of these, and of investigations by others, would have demonstrated conclusively that, apart from the figure after the decimal point, the atomic weight of tellurium was definitely higher than that of iodine, and that tellurium was to be considered as an exceptional element in the periodic system. In 1907, however, Marckwald announced a different figure, namely, 126·85, which he obtained by converting telluric acid, purified by several hundred crystallizations, into tellurium dioxide. This value is slightly lower than the figure then accepted for iodine, namely, 126·95, and great interest was accordingly aroused. Criticism was chiefly directed against Marckwald as regards the choice of telluric acid as a suitable substance for work of extreme precision; in particular, it was urged that the crystals might contain water mechanically enclosed. In a subsequent research Marckwald and Foizik abandoned telluric acid and revived a volumetric oxidation method which had been used by Brauner. The value which they obtained in 1910 by this latter method was 127·6.

The conclusions reached by all the most reliable subsequent

investigations on the atomic weight of this element are that the atomic weight of tellurium is about 0·7 of a unit higher than that of iodine, and that the element can be obtained in a spectroscopically pure condition. There is no evidence that the cause of the high atomic weight is to be explained by the presence of a non-separable dwi-tellurium such as Mendeleeff supposed. There is an element belonging to this group in the periodic table, namely, the radioactive element polonium, discovered by Madame Curie and shown by Marckwald to be an analogue of tellurium, with an atomic weight of 210 derived from a study of its radioactive properties, but it is unstable, its half-period is about 140 days, and it is quite impossible that the extremely minute quantities in which it is available can have had the slightest influence upon the atomic weight of tellurium.

It had been noted by Mendeleeff and by others that the electrochemical character of the elements is well displayed by their position in the several groups in the table. Thus the strongly electropositive nature of the alkali metals in Group I shows a gradual increase with increase in atomic weight as we pass from lithium to caesium. Turning to Group VII we see electronegative properties diminish in intensity as the atomic weight increases from fluorine to iodine. Viewing the table as a whole we see a gradual diminution of electropositive properties as we pass from Group I to Group VII. These considerations must have received much attention in connexion with the properties of argon and helium. After the fundamental work of Rayleigh and Ramsay in 1894–5 on the discovery of argon, followed by work on helium by Ramsay in 1898, the existence of the other three gases, now known as neon, krypton and xenon, was predicted and subsequently verified experimentally by Ramsay and Travers. To give expression to their chemical inertness they were placed in a new Group 0 at the extreme left of the table to demonstrate their zero valency.

It is instructive to compare the early form of periodic table with that which was published in the third English edition of Mendeleeff's *Principles of Chemistry* (1905). The number of elements in the table was increased from 63 to 71, and the newly discovered element radium was placed in Group II in the 12th series. The tervalency of lanthanum was recognized by placing that element in Group III, with its atomic weight determined as 139 after barium in Group II, and 'didymium' was removed from the table. The substance previously considered to be an element was shown by Auer von Welsbach in 1885 to consist of two elements, namely, praseodymium having green salts, and neodymium having rose-coloured salts. The absorption spectra are highly characteristic. The elements scandium and gallium were placed in Group III under boron and aluminium respectively, and germanium was placed in Group IV under silicon. Both in the original table and in the revised one the metals copper, silver, and gold were placed in Group I and also in Group VIII along with the transitional elements, but whereas in the 1869 table brackets were placed around the coinage metals in Group I, the brackets were transferred to them in Group VIII in the revised table, which appears to indicate that Mendeleeff must have considered that their univalent character was clearly defined.

The position of the periodic law at the beginning of the twentieth century was recognized as a far-reaching generalization, which had been the agency whereby the existence of new elements and their properties had been predicted and subsequently verified, and doubtful or erroneous atomic weights had been checked or corrected. But some grave defects still remained, notably the inversion of the order in the cases of tellurium and iodine, of argon and potassium, and of nickel and cobalt. Attempts to fit the rare earths into the table had achieved a very limited degree of success. These and other defects in the system remained unsolved until it was shown (Chapter v) that the fundamental properties of an element are determined not by its atomic weight but by another constant known as its atomic number. It will be seen later that the periodic law has acquired increased importance when the elements are considered from the standpoint of their atomic numbers, and that all anomalies and irregularities have disappeared.

Reference has already been made to the contributions of Werner and of Abegg to the theory of valency, and a very brief description of their theories must now be given. In 1891 Werner drew a classificatory distinction between what he termed *principal* or *primary* valencies, and *auxiliary* or *secondary* valencies, the former being particularly concerned with the ionizable parts of molecules and the latter with those parts of molecules which are non-ionizable. He also introduced a conception which has since become of great importance, namely, the co-ordination number. Without making any assumptions regarding the nature of atoms or of chemical combination, apart from the conception of ionization, Werner postulated that auxiliary valencies were capable of binding not only atoms, but also molecules which can have an independent existence such as those of water and ammonia. These ideas were applied with conspicuous success to the discussion of isomerism in inorganic compounds, the number and complexity of which is considerable. Thus metals such as cobalt and chromium can give rise to salts, which can form hydrates and also additive compounds with ammonia, and were formulated by Werner in a way which enabled their distinctive reactions to be expressed. The names formerly given to these classes of compounds, such as luteo-cobaltic chloride, purpureocobaltic chloride, and roseocobaltic chloride, were expressed by Werner as hexammine cobaltic chloride, $[(NH_3)_6Co]\overset{+\,+\,+}{}\,\overset{-\,-\,-}{Cl_3}$, chloropentammine cobaltic chloride, $[Cl(NH_3)_5Co]\overset{+\,+}{}\,\overset{-\,-}{Cl_2}$, and aquopentammine cobaltic chloride, $[(H_2O)(NH_3)_5Co]\overset{+\,+\,+}{}\,\overset{-\,-\,-}{Cl_3}$. In these compounds the co-ordination number is six, that is, the total number of chlorine atoms and ammonia or water molecules around the metallic atom within the square brackets. It will be noted that in the hexammine and in the aquopentammine salts the cation is tervalent, but in the chloropentammine salt the valency of the cation has fallen by one unit. Compounds of the type $[Cl_3(NH_3)_3Co]$ are well known and are non-electrolytes. Co-ordinated metallic compounds in which the principal metallic constituent is in the acidic ion are well exemplified by compounds such as potassium ferrocyanide, $\overset{+\,+\,+\,+}{K_4}\,[\overset{-\,-\,-\,-}{Fe(CN)_6}]$, and also potassium ferricyanide, $\overset{+\,+\,+}{K_3}\,[\overset{-\,-\,-}{Fe(CN)_6}]$. The co-ordination number six is certainly the most common one, but compounds with the co-ordination numbers of four and occasionally of eight are also well known.

Werner's theoretical ideas were by no means limited to the discussion of structural isomerism among inorganic compounds, but were extended to stereochemical considerations. And in more recent times much attention has been given to them by Sidgwick and others in connexion with the development of electronic theories of valency.

The problems of valency were approached from a somewhat different standpoint by Abegg and Bodländer in 1899, who employed a system of chemical classification based upon electro-affinity. They pointed out that elements of weak electro-affinity display a strong tendency to form complex ions. In 1904 Abegg developed these ideas further and put forward a theory according to which every element is endowed with two kinds of valency, viz. its normal (positive) valency and what he termed its (negative) contravalency. The sum of the normal and contra-valencies is equal to eight. It may be remarked that in some respects Abegg really made very little advance upon Mendeleeff, who put forward similar considerations at the time that he drew up his periodic table in 1869, but with this difference that Abegg laid great stress upon the electrochemical aspects of his theory. Another way of regarding Abegg's theory of valency is to consider it as a revival of the old dualistic electrochemical ideas of Berzelius welded on to the periodic classification. The significance of the theory can be readily understood by reference to a few simple examples. Thus phosphorus has a normal valency of three in phosphine and a contravalency of five in phosphorus pentachloride, the sum of which is eight. Similarly, the sum of the two kinds of valency of sulphur is eight, the normal valency being two in hydrogen sulphide, and the contra-valency six in sulphur hexafluoride. But with many elements this simple relation cannot be seen, and in such cases Abegg got over the difficulty by assuming that the contravalencies were latent. Still more difficult to reconcile with Abegg's rule of eight is the valency of boron. In most of its compounds, and particularly in the halides, the element is undoubtedly tervalent. But in the hydrides, which have been very carefully studied by Stock since 1912, it would appear that the element is quadri-valent. Altogether Abegg's ideas have been very much less fruitful in the later developments of chemistry than those of

Werner, and are now only of historic interest, but it has been suggested that they may possibly have been of some influence in the development of the more recent octet theory of atomic structure.

In 1899, Thiele brought forward what he termed a theory of partial valencies, the object of which was to discuss some of the peculiarities of compounds having conjugated double bonds in the molecule, particularly with reference to reactions involving addition. Thus when the molecule of butadiene

$$H_2C=CH-CH=CH_2$$

adds two atoms of bromine, the dibromide which is formed has the bromine atoms attached to each of the two terminal carbon atoms, and simultaneously a new double bond appears thus,

$$BrH_2C-CH=CH-CH_2Br.$$

In compounds of this kind Thiele regarded the carbon atoms as possessing a certain residual affinity—a term variously used to denote latent combining power—which he designated as their partial valencies, and he wrote the formula of butadiene as

$$H_2C^{\cdots}CH-CH^{\cdots}CH_2$$

to give expression to the way in which it reacts by addition. In such a conjugated system the partial valencies of the middle pair of carbon atoms can unite and become inactive, thus leaving those of the two end carbon atoms free to combine. The bonds between the middle pair then become an ordinary double bond. Thiele extended his ideas to aromatic compounds with the object of giving expression to the distinctive features of their unsaturated character. Up to 1899 the two rival formulae for benzene were the Kekulé formula with oscillating double bonds, and the centric formula, first suggested by Armstrong in 1887 and later by Baeyer in connexion with the dihydroterephthalic acids. What Thiele did was to modify the Kekulé formula by interpolating his partial valencies within the double bonds, but it cannot be said that any fundamental advance resulted from this improvement of the formula.

Chemistry at the close of the nineteenth century bore all the marks of a highly flourishing science, and it would seem right to attribute much of this success to the atomic theory. Nevertheless, it cannot be denied that much progress was made in various directions, while theoretical ideas were still in a confused state, before Cannizzaro's promulgation of the necessity of accepting Avogadro's teaching. Thus inorganic analysis associated with the names of Berzelius, Gay-Lussac, Stas, and Bunsen, had reached a high standard of accuracy, and some of the less common elements were discovered during the earlier years of the century. But it must be borne in mind that much progress in inorganic chemistry could be made without the necessity of using modern formulae. With organic chemistry the state of affairs was rather different. Until molecular formulae based upon Avogadro's rule were in general use, it was extremely difficult to enter into problems of constitution with any great prospect of success. It would be right to say that the highly developed system of organic chemistry has been largely built upon the foundations laid by Frankland and by Kekulé. It should be noted that problems of constitution, and, be it added, of configuration, of highly complicated molecules were solved successfully without entering into such difficult problems as the structure of atoms or making assumptions regarding the nature of valency.

It might have been supposed that atomic weight research which was such a strong feature of nineteenth-century chemistry would have disposed of Prout's hypothesis, since it was abundantly clear that the values of the atomic weights were not whole numbers, but that the figures after the decimal point had a real significance. Such, however, was not the case. A considerable number of chemists were of opinion that some modification of Prout's ideas would eventually be forthcoming. Indeed, it might be said that there were two schools of thought, one of which favoured the view that the elements were all derivatives of some fundamental substance, and the other, to which Mendeleeff belonged, which opposed such a view. It is interesting to recall that Crookes's investigations on the rare earths led him in 1889 to favour the idea of some fundamental substance. His views were strongly combated by others who worked in this field, such as Marignac and Lecoq de Boisbaudran.

Regarding the nature of atoms, chemists were for the most part wisely reticent. Many were content to regard the atomic theory as a useful instrument which led to important advances in chemistry, but were wholly agnostic regarding the nature and even the physical reality of atoms and molecules. On this subject the views of Faraday in a letter dated 2 August 1853 to Henry, the biographer of Dalton, are of particular interest: 'I do not know that I am unorthodox, as respects the atomic hypothesis. *I believe in matter and its atoms* as freely as most people, at least I think so. As to the little solid particles, which are supposed by some to exist independent of the forces of matter...as I cannot form any idea of them apart from the forces, so I neither admit nor deny them. They do not afford me the least help in my endeavour to form an idea of a particle of matter. On the contrary they greatly embarrass me....' Henry regarded Faraday's views on the constitution of matter as remarkably similar to those of Father Boscovich, who as long ago as the year 1763 considered atoms as mathematical points endowed with the property of inertia and with mutual attractions and repulsions depending upon the distances between them. The views of Boscovich have received attention from modern physicists, e.g. Lord Kelvin (1889) and later by J. J. Thomson, who found them useful in the early stages of the evolution of his views on atomic structure (1907).

This latent scepticism regarding the physical reality of atoms and molecules survived into the early years of the twentieth century. Thus Ostwald, whose writings were of considerable influence on the study of chemistry, had a strong dislike of mechanical theories of any kind and particularly of the kinetic theory of gases, and thus constantly refrained from using the ordinary terminology, e.g. *atomic weight* and *molecular weight*, but preferred to use expressions such as *combining weight* and *molar weight* instead. Although the experimental evidence in favour of the physical reality of atoms and molecules was much less plentiful then than at the present time, it was nevertheless fairly considerable. In this connexion it is sufficient to advert to Lord Rayleigh's views (1871) on the blue colour of the sky as being due to diffraction of light by particles of molecular dimensions, and to the evidence forthcoming from van der Waals's equation (1873). It was therefore with considerable surprise that chemists received Ostwald's Faraday Lecture delivered before the Chemical Society in 1904, in which from the principles of chemical dynamics he sought to establish the fundamental stoichiometrical laws without the aid of the atomic theory. Ostwald considered that the principles of energetics provided a more substantial foundation for the science than any theory based upon the assumption of the physical reality of atoms and molecules. In his Faraday Lecture, Ostwald elaborated certain fundamental ideas due to Willard Gibbs (1874-8), and gave support to views which had been expressed about 1895 by Wald who considered that the atomic theory could be abandoned. He then attempted to define conceptions such as those of element, compound, chemical individual, and to establish the laws of combination. The reception of these views might be described as one of mixed bewilderment and suspicion. In any case Ostwald's views secured few if any adherents, and strong objection was raised by some regarding the soundness of the reasoning involved. It is obviously extremely difficult to understand how questions of constitution could be discussed without making use of the atomic theory, and it is difficult to imagine a more inopportune time for such views to have been expressed when the study of atomic structure was in the ascendant, and actually in the very year after Ramsay and Soddy announced the transformation of radium emanation into helium, one of the early triumphs of the theory of atomic disintegration.

REFERENCES

Sir Edward Frankland. *Experimental Researches in Pure, Applied and Physical Chemistry*. London, 1877.

A. W. Williamson. *Papers on Etherification and on the Constitution of Salts*. Alembic Club Reprints, No. 16.

S. Cannizzaro. *Sketch of a Course of Chemical Philosophy*. Alembic Club Reprints, No. 18.

A. S. Couper. *On a New Chemical Theory and Researches on Salicylic Acid*. Alembic Club Reprints, No. 21.

A. N. Meldrum. *Avogadro and Dalton. The Standing in Chemistry of their Hypotheses*. Aberdeen University Studies, No. 10. 1904.

A. Werner. *New Ideas on Inorganic Chemistry*. Translated from the second German edition by E. P. Hedley. London, 1911.

F. Penny. *Phil. Trans.* 1839, pp. 13 *et seq.*

J. S. Stas. *Œuvres Complètes*. Bruxelles, 1894.

J. C. Galissard de Marignac. *Œuvres Complètes.* Genève, 1902.

T. W. Richards. This author has written a number of papers on atomic weight determinations in the *Publications of the Carnegie Institution of Washington* between 1905 and 1910.

Lecoq de Boisbaudran and de Lapparent. The Telluric Screw. *Chem. News*, 1891, LXIII, 51.

T. M. Lowry. Dynamic Isomerism. *Report British Association*, 1904, p. 193.

W. Ostwald. Faraday Lecture. Elements and Compounds. *J. Chem. Soc.* 1904, p. 506.

R. Abegg. *Z. anorg. Chem.* 1904, XXXIX, 343.

[References to the literature on atomic weight determinations of the elements will be found collected together in the volumes of J. W. Mellor's *Comprehensive Treatise on Inorganic and Theoretical Chemistry*, and in Abegg's *Handbuch der anorganischen Chemie.*]

C. Schorlemmer. *The Rise and Development of Organic Chemistry.* Revised edition by A. Smithells. London, 1894.

H. E. Roscoe and A. Harden. *A New View of the Origin of Dalton's Atomic Theory.* London, 1896.

Lothar Meyer. *Modern Theories of Chemistry.* Translated from the fifth German edition by P. Phillips Bedson and W. Carleton Williams. London, 1888.

D. Mendeleeff. *The Principles of Chemistry.* Third English edition. Translated from the Russian seventh edition by G. Kamensky and edited by T. H. Pope. Two volumes. London, 1905.

H. V. Gill, S.J. *Roger Boscovich, S.J. (1711–87), Forerunner of Modern Physical Theories.* Dublin, 1941.

Chapter II

ELECTROCHEMISTRY

Although the beginnings of electrochemistry are commonly associated with the distinguished names of Davy and Faraday, it should not be forgotten that their far-reaching discoveries were brought about in some degree as the result of the work of earlier investigators. Some of the earliest electrochemical experiments were bound up with the problem of the identity of frictional and voltaic electricity. In 1800 Nicholson and Carlisle, using a voltaic pile as their source of current, found that if platinum wires connected with the terminals were placed in water there was an evolution of hydrogen from one wire and of oxygen from the other. They explained the phenomenon correctly as a decomposition of water. In the following year the identity of static and current electricity was claimed to have been established by Wollaston, who found that water could be decomposed by either kind of electric energy; but his proof was not considered to be altogether convincing, and the subject was finally settled by Faraday some thirty years later.

The decomposition of water by Nicholson and Carlisle greatly interested Davy who commenced a series of electrochemical experiments, which culminated in the isolation of the metals of the alkalis in 1807. These small beginnings may be described as the foundation of the modern electrochemical and electro-metallurgical industries, since many substances are now produced by methods which are based on these early discoveries.

It was observed by the first experimenters that the products of decomposition always appeared close to the wires which carried the current, and never throughout the body of the liquid. In 1806 an attempt was made by Grotthus to account for this. He considered that the decomposition was brought about as the result of the attractive and repulsive forces of the electricity delivered by the poles of the battery. The views of Grotthus were well summarized by Faraday who said that 'the pole from whence resinous electricity issues attracts hydrogen and repels oxygen, whilst that from which vitreous electricity proceeds attracts oxygen and repels hydrogen; so that each of the elements of a particle of water, for instance, is subject to an attractive and a repulsive force acting in contrary directions, the centres of action of which are reciprocally opposed.... He explains the appearance of the elements at a distance from each other by referring to a succession of decompositions and re-compositions occurring amongst the intervening particles....'

Other contemporary investigators, e.g. de la Rive, put forward alternative theories of the chemical decomposition, but they amounted to much the same idea, namely, decomposition by the electric current and transference of the products to the poles. It may be remarked that little progress was made towards the establishment of any sort of consistent or satisfactory theory until after the quantitative researches of Faraday. As regards the *qualitative* aspect of the subject, however, the recognition of the nature of the substances liberated at the positive and negative poles immersed in a liquid resulted in the promulgation in 1812 by Berzelius of his celebrated dualistic theory of chemical combination, according to which salts are to be regarded as derived from the union of a basic (electropositive) oxide with an acidic (electronegative) oxide.

Very important and far-reaching progress in electrochemical research was made by Faraday in 1833 and 1834. In the former year he enunciated his two well-known laws regarding the relation between the strength of a current and the amount of chemical decomposition effected by it, and of the proportionality between the masses and the equivalent weights of different substances liberated when the same current is passed through a series of solutions. Throughout this work Faraday showed himself a very original and skilful experimenter, and he was anxious to secure that the results of his researches should be expressed in the most suitable language. Accordingly, he carried on a correspondence with Whewell in 1834 regarding terminology, which resulted in the introduction of the now well-known terms *electrode, electrolysis, electrolyte, ion, anode,* and *cathode* into chemistry. Faraday's remarks on the word *ion* are perhaps worth quoting: 'Finally, I require a term to express those bodies which can pass to the *electrodes*, or, as they are usually called, the poles. Substances are frequently spoken of as being *electro-negative* or *electropositive*, according as they go under the supposed influence of a direct attraction to the positive or negative pole. But these terms are much too significant for the use to which I should have to put them; for though the meanings are perhaps right, they are only hypothetical, and may be wrong; and then through a very imperceptible, but still very dangerous, because continual, influence, they do great injury to science, by contracting and limiting the habitual views of those engaged in pursuing it. I propose to distinguish such bodies by calling those *anions* which go to the *anode* of the decomposing body; and those passing to the *cathode*, *cations*; and when I have occasion to speak of these together, I shall call them *ions*.'

The enunciation of Faraday's laws of electrolysis exercised a profound influence upon the subsequent development of electrochemical research. Indeed, it is difficult to think of any of the more modern quantitative developments of the subject which have not been in some way influenced by Faraday's work. Of outstanding importance is the recognition of the constant value, appropriately named the *faraday*, of the charge carried by a univalent ion in electrolysis. Helmholtz in his Faraday Lecture in 1881 pointed out that a direct consequence of the principles of electrochemical equivalence must be the development of some form of atomic theory of electricity. Another and more practical aspect of the importance of electrochemical equivalents of the elements has been the means of defining certain electric quantities. Thus in 1908 an International Conference in London defined the ampere as the current which when passed through a solution of silver nitrate under certain specified conditions deposits silver at the rate of 0·001118 g. per sec.

Some experiments carried out by Daniell, the inventor of the well-known cell of constant electromotive force, between 1839 and 1844, may be described as the first beginnings of a correct understanding of the mechanism of the electrolytic processes. He showed that the ions of a solution of a salt of an oxy-acid were not a positively charged basic *oxide* as the cation and a negatively charged acidic *oxide* as the anion, according to the dualistic theory of Berzelius, but that the cation and anion consist of the ions of the *metallic* and *acidic* constituents of the salt. Moreover, he showed that when a solution of a salt such as sodium sulphate is electrolyzed the hydrogen and oxygen liberated at the electrodes are secondary products of the reaction. Daniell also showed that the liberation of the products of electrolysis at the two electrodes in equivalent proportions is by no means necessarily due to equality in the rates of movement of the cation and anion towards their respective electrodes. Some experiments which he conducted in conjunction with Miller, using a porous diaphragm to separate the anode and cathode liquids, provided direct evidence of inequality in the rates of transference of the ions in the solutions although there was strict equivalence in the ions liberated at the electrodes.

Between 1853 and 1859 a long and elaborate series of experiments on the migration of ions were carried out by Hittorf. The general method of carrying out these experiments was to conduct the electrolysis of the salts under investigation in cells which permitted the liquid around the anode and the cathode to be removed for analysis. Hittorf employed Grove's cells as his source of current, and the electrolysis was always conducted in series with a voltameter, usually a silver voltameter, to measure the total quantity of electricity which had been transmitted through the circuit. The results of these experiments were expressed as what Hittorf termed *transport numbers*, or the ratio of the amount of salt taken from the vicinity of one of the electrodes to the total amount which disappeared. Hittorf was able to show that the values of the transport numbers were, within a wide range, independent of the strength of the current, but varied to some extent with the concentration of the solutions. This variation of the value of transport numbers with change of concentration has been traced in many cases to the existence of complex ions in the more concentrated solutions, the complexes being broken down by dilution; in other cases there is little doubt that it is due to a dragging of the solvent along with the ions. Hittorf showed that when solutions of salts such as potassium argenticyanide, potassium platinichloride, potassium ferrocyanide, and potassium ferricyanide were electrolyzed the heavy metal in every instance migrated towards the anode. He was thus able to establish the fundamental distinction between double salts and complex single salts. He also showed that complex ions differ greatly in stability, and that therefore the distinction between the two classes of salts was perfectly definite in extreme cases. In other cases the distinction is one of degree rather than fundamental, as Hittorf was able to show with potassium cadmium iodide, a salt which in concentrated solution was found to display the properties of a complex salt, but in dilute solution the complex ions were found to break down.

The numerical values for transport numbers were found to be proper fractions, the sum of the values for the two ions being equal to unity for all salts in which there were no complications arising from the existence of complex ions. On the other hand, a value for a transport number numerically greater than unity is direct evidence of the formation of complex ions. Somewhat more difficult to understand, however, were the relatively small variations in the value of transport numbers with change of concentration of the electrolyte in which the occurrence of complex formation was most unlikely to arise. This problem occupied the attention of numerous investigators since Hittorf's time, and has been solved chiefly as the result of experiments carried out by Washburn and others since 1908. Washburn employed Hittorf's methods, modified by adding a suitable non-electrolyte as a reference substance, so as to disentangle the motion of the ions from that of the solvent during the electrolysis. He found that the differences between the values of the Hittorf and the 'true' transport numbers gradually disappeared with increasing dilution of the solutions.

A series of measurements on the electrical conductivity of aqueous solutions carried out between 1869 and 1879 by Kohlrausch, taken in conjunction with Hittorf's work, provided a means of calculating the mobility of the ions, and incidentally of obtaining a clearer understanding of the mechanism of the electrolytic processes. It may be noted that Ohm's well-known law expressing the relation between the strength of a current flowing through a conductor and the electromotive force applied to it was enunciated as early as the year 1827, but its importance was not recognized at the time. As regards electrolytes there was considerable uncertainty for many years as to how far the law was applicable. Eventually it was shown by Kohlrausch that, provided the effects of polarization are avoided, Ohm's law is strictly obeyed. Satisfactory measurements of the conductivity of solutions of electrolytes were made by using alternating currents in order to eliminate the effects due to polarization. As the result of numerous experiments Kohlrausch in 1879 was able to establish the principle that at infinite dilution the equivalent conductivity of a salt can be represented as the sum of the conductivities due to the cation and the anion. This important generalization was established by studies on the conductivity of mixed electrolytes. For example, Kohlrausch showed that the conductivity of a solution of sodium chloride and potassium nitrate was equal to the conductivity of an equivalent solution of sodium nitrate and potassium chloride. By considering the separate values for the four salts Kohlrausch was able to obtain values for the mobilities of the ions. He found that when a solution of an electrolyte was progressively diluted the value of the molecular or equivalent conductivity increased with dilution, and eventually reached a constant value, known as the value at infinite dilution. The values for potassium chloride and potassium nitrate were found to be 130·0 and 126·3 respectively, and hence the difference between the values for the chloride and nitrate ions was 3·7 units. For sodium chloride and sodium nitrate the equivalent conductivity values were 108·9 and 105·2 respectively, showing the same difference, viz. 3·7 units as before. The difference for the potassium ion and the sodium ion can be obtained by subtracting the values for the two chlorides, viz. 130·0 − 108·9, or the two nitrates, viz. 126·3 − 105·2, giving a result of 21·1 units in each case. Studies arising out of the discovery of the law of the independent mobility of the ions continued to engage the attention of Kohlrausch for over thirty years after the principle was first enunciated. Incidentally, in conjunction with Heydweiller, Kohlrausch in 1894 accomplished an operation of great experimental difficulty by preparing what was the nearest approximation to pure water and showed that it possessed a small, but definite, conductivity, which must be attributed to the water itself and not to dissolved impurities. It was also shown by Kohlrausch that the hydrogen ion has an exceptionally high degree of mobility, several times as great as that of other ions. This high value for the mobility of the hydrogen ion has an important bearing upon the comparison of the strengths of acids by measurement of the electric conductivity.

It has been remarked that the first stage in the study of transport numbers is associated with the name of Hittorf, and the second with that of Kohlrausch. What may be called the third stage, the measurement of ionic velocities by a moving boundary method, was begun in 1886 by Lodge and afterwards improved by Whetham (1893–5), Masson (1899), Steele (1902) and by MacInnes and his collaborators (1923–7). Lodge made use of solutions of electrolytes dispersed in jellies and employed colour devices to indicate the movement of the ions. Thus the velocity of hydrogen ions was determined by using a jelly containing faintly alkaline sodium chloride and coloured red with phenolphthalein. The mixture was placed in a horizontal glass tube connected with two vessels filled with dilute sulphuric acid. On passing a current from one vessel to the other through the tube the red colour of the alkaline phenolphthalein was gradually decolorized as the hydrogen ions migrated from the anode vessel through the tube. Lodge was thus able to measure the velocity of the hydrogen ions through the jelly under a known potential gradient. His results were in fundamental agreement with those of Kohlrausch as calculated from the conductivity measurements.

Whetham devised an elegant method for determining ionic velocities. He placed two solutions, such as potassium dichromate and potassium carbonate, one above the other in a vertical glass tube. The solutions were as far as possible of equal specific resistance, of equivalent concentration, of different density and of different colour. On passing a current through the solution the velocity of the ions could be determined by measuring the rate at which the colour boundary moved. The results obtained by Whetham were, in general, in good agreement with those calculated from Kohlrausch's theory. Further studies on moving boundaries were made by Masson who used cupric salts as a source of coloured cations, and potassium chromate to provide coloured anions, and eventually devised a highly accurate method for determining ionic velocities. Masson's method was further elaborated by Steele, who found that the migration of colourless ions could be followed by observations of differences in the refractive index across the boundary. Apart from their value for measuring ionic velocities, these visual methods have provided definite proof of the independent mobility of the ions, and incidentally have been of much value for recognizing complex ions.

As soon as it became recognized that electrolytic conduction followed Ohm's law, the deficiencies in the Grotthus chain theory of electrolysis became apparent. Clausius in 1857, realizing that no electrical work was expended in decomposing the molecules of the electrolyte into ions, followed up an earlier idea due to Williamson and considered that a small fraction of the molecules of an electrolyte was dissociated into ions at the outset. It was therefore possible to consider the process of electrolysis as a kind of convection, the ions moving through the solution and carrying their charges with them. Thirty years later, in 1887, Arrhenius put forward his theory of electrolytic dissociation, which differed from that of Clausius as regards the *extent* of the dissociation. Whereas Clausius regarded the degree of dissociation to be very small, Arrhenius regarded it as considerable, and in the case of strong electrolytes in dilute solution to be almost complete. In formulating his theory Arrhenius took careful account of van't Hoff's theory of dilute solutions, and particularly of the abnormally high osmotic properties of electrolytes. These phenomena were correlated with the conductivity experiments of Kohlrausch. Of particular interest were the respective contributions of van't Hoff and of Arrhenius to the calculation of the degree of ionization of an electrolyte. Van't Hoff published his theory of dilute solutions, in which he stressed the analogy between osmotic pressure and gaseous pressure, shortly before the appearance of Arrhenius's theory of ionization. But it is significant that in his paper on osmotic pressure, van't Hoff remarked that he would not have attempted to extend Avogadro's theorem to electrolytes had not Arrhenius written to him privately pointing out the probability that substances of a saline character were dissociated into ions. The celebrated coefficient i, which was introduced by van't Hoff to express the ratio of the actual osmotic effect produced by an electrolyte to the effect which would be produced if it behaved like a non-electrolyte, at once attracted the attention of Arrhenius. In calculating what is known on the classical theory as the degree of ionization, Arrhenius used a term which he defined as the activity coefficient, namely, the ratio of the number of active molecules to the total number of molecules, active and inactive. It may be noted that the term *activity coefficient* has reappared in more modern physical chemistry, but in a different sense as originally understood by Arrhenius. According to Arrhenius the degree of electrolytic dissociation of a substance in solution can be calculated from the ratio of the equivalent conductivity at the dilution under consideration to that at infinite dilution, λ_v/λ_∞ and also by van't Hoff's method from osmotic pressure measurements, or from properties proportional to these, such as depressions of the freezing-points of solutions. Approximate, but not exact, agreement between the two methods was realized for a large number of acids, bases, and salts. The subsequent history of the ionic theory has been very much concerned with difficulties connected with obtaining more exact agreement between different methods for determining degrees of ionization.

When the theory of electrolytic dissociation was first promulgated it met with a very mixed reception, favourable for the most part, but nevertheless accompanied with bitter hostility from some chemists. In attempting to view this subject in retrospect it must be remembered that Arrhenius was singularly fortunate in having so influential a champion of the new ideas as Ostwald, who for twenty years directed experimental work with the object of placing the theory in the forefront of practically all chemical research. A certain number of chemists 'resented' the theory without putting forward anything better by way of an alternative. Admittedly the theory enabled a large and varied collection of phenomena to receive explanation, and to a considerable extent in a quantitative manner. The most serious objection to the theory, and one which was scarcely ever seriously faced at that time, was the problem of the energy required to sever the molecules of admittedly stable substances into their ions. FitzGerald in his Helmholtz Memorial Lecture in 1896 criticized the views of van't Hoff and Arrhenius very seriously. He began by pointing out that although there was a close parallelism between the laws of gaseous pressure and of osmotic pressure, it by no means followed that the physical processes were of the same nature, and added significantly that 'the dynamical condition of molecules in solution is essentially and utterly different from that of a molecule in a gas'. Similar views were expressed on more than one occasion by Kelvin. The exceptional capacity of water, as compared with other solvents, for ionizing substances dissolved in it was explained by J. J. Thomson and independently by Nernst in 1893 as due to the very high specific inductive capacity of the liquid. They pointed out that if the forces holding the ions together in a molecule were electrical in origin, they would be greatly weakened by immersing the molecule in a medium of high specific inductive capacity. In 1894 it was shown by Whetham that there is a considerable parallelism between the conductivities of many salts dissolved in water, methyl alcohol, and ethyl alcohol and the dielectric constants of these liquids. Other investigators have drawn attention to the same sort of parallelism with certain other solvents. The importance of a high specific inductive capacity of a liquid and its power of ionizing substances dissolved in it was recognized by FitzGerald, but he was very careful to add that 'all this hangs together, but it lends no support at all to the dynamically impossible theory that the ions are *free*'. This problem continued to be a source of difficulty until the crystalline structure of salts was examined by X-ray analysis, when it was found that dissolved salts make good electrolytes and have ionic lattices in the crystalline state. As the ions pre-existed in the crystals it was therefore no longer essential to look for a source of energy necessary to sever the molecules into ions.

A great deal of work has been done on comparisons between the degree of ionization as determined from experiments on the freezing-points of electrolytes and from measurements of the conductivities. In making such comparisons it is essential for the conductivities to be determined at temperatures close to the freezing-points. Thus Whetham in 1900 made some determinations of the conductivities of highly dilute solutions of potassium chloride for comparison with freezing-point determinations by Griffiths, who employed the most exact methods of platinum-resistance thermometry. Differences of the order of 2 or 3 per cent were invariably found in the values of the degree of ionization as determined by the two methods, those obtained from measurements of the conductivity being always the higher. Explanations of these differences were not forthcoming at the time, but are now better understood in terms of developments of the modern theory of the complete ionization of strong electrolytes.

Another source of difficulty with the original theory of electrolytic dissociation was the behaviour of strong electrolytes with respect to the law of mass action. In 1888 Ostwald assumed that the simple laws relating to the thermal dissociation of gaseous molecules could be extended to the reversible ionization of the molecules of an electrolyte into its ions. He expressed his reasoning in the form of an equation known as Ostwald's dilution law, according to which the molecular conductivity should be proportional to the square root of the dilution. For weak electrolytes, such as acetic acid, the accuracy of the law has been verified with great strictness, as was actually shown from the conductivity measurements of Kohlrausch as early as 1878. For strong electrolytes the law does not hold even approximately. It has been remarked by some that it is really not surprising that the law of mass action should fail for strong electrolytes, since the solutions would require to be at extreme dilution for interionic forces to cease being operative.

Studies on the effect of dilution on the equivalent conductivity of the alkali metal salts of different acids led Ostwald in 1887 to the formulation of a valuable, albeit empirical, rule regarding the basicity of the acids. The differences between the equivalent conductivities of the sodium salts of acids at dilutions of 32 and of 1024 litres were found to approximate to 10 units for a monobasic acid, 20 units for a dibasic acid, 30 units for a tribasic acid and so on. The practical value of the rule has been amply demonstrated since that time.

The limitations of the classical theory of ionization, admittedly an extremely useful theory from various points of view, engaged the attention of several investigators, particularly of Bjerrum (1906–10), of Hantzsch (1906), of Milner (1913) and in more elaborated detail by Debye and Hückel (1923). All were agreed in considering strong electrolytes to be completely ionized at all concentrations. Bjerrum and Hantzsch were led to their views regarding total ionization chiefly as a result of studies on the optical behaviour of coloured electrolyte solutions. They found that in many cases the light absorbed by a given amount of a solute was independent of the concentration. Hence it at once followed that, unless the undissociated molecules absorb light in exactly the same way as the ions, the degree of dissociation must be independent of the concentration. Bjerrum regarded solutions of weak electrolytes to contain undissociated molecules, but none to be present in solutions of strong electrolytes. Since the modern theory has been formulated it has received very strong support from the results of X-ray studies on the crystals of electrolytes.

In 1907 Bjerrum introduced an osmotic coefficient as a measure of the deviation of the osmotic pressure of real solutions from the ideal value, and in the same year Lewis introduced an activity coefficient to replace the values of the Arrhenius degree of dissociation as determined from the ratio λ_v/λ_∞. The purpose of these various coefficients, now known generally as deviation coefficients, is to introduce corrections into the mathematical equations so as to give expression to the *effective* as distinct from the *actual* concentration of the ions. It should be remembered that van't Hoff's coefficient i has the same *practical* meaning as it had originally, namely, the ratio of the actual osmotic effect produced to the value calculated for this osmotic effect if the solute behaved as a non-electrolyte. At the present time it would be correct to call the ratio λ_v/λ_∞ the *apparent* degree of ionization. The calculation of the degree of ionization from van't Hoff's coefficient i can no longer be regarded as correct without introducing various factors.

In an introduction to Falkenhagen's work on electrolytes, Debye in 1932 stated that the relation of the modern theory of complete ionization to the classical theory of Arrhenius might be compared with the relation of van der Waals's theory of gases to the simple gas laws which are related to the theory of perfect gases. 'The modern theory of electrolytes is not incompatible with the ideas of Arrhenius.' The classical theory can certainly be retained for weak electrolytes. A more pessimistic view regarding the whole position of the theory of electrolytic dissociation was expressed by Bancroft in the Golden Jubilee Number of the *Journal of the American Chemical Society* in 1926. 'It must be admitted that the electrolytic dissociation theory has suffered much more from its friends than from its foes. Forty years of intensive development have brought us to the point where we cannot determine any electrolytic dissociation with any degree of accuracy, and where we question the significance of the term electrolytic dissociation.'

Although it is now recognized that the *approximate* agreement between the results obtained by determinations of van't Hoff's coefficient i and by measurements of the electrical conductivity of strong electrolytes can no longer be used in support of the classical theory of ionization, but requires the refinements of the theory of Debye and Hückel for its discussion, the results have nevertheless been of great practical value in obtaining information regarding the constitution of electrolytes in solution. Determinations of i by measurement of the freezing-point have been found of much value in supplementing information derived from the measurements of conductivity. Thus values of i within certain limits indicate the number of ions present in solution and can thus confirm conclusions drawn from differences in the equivalent conductivities between specified limits. Some of the earlier investigators assigned the molecular formula $K_2Mn_2O_8$ to potassium permanganate. Bredig in 1893 showed that the differences between the electrical conductivities for one gramme-equivalent in 32 and in 1024 litres was 10·7 units, thus showing clearly the monobasic character of permanganic acid. Determinations of the freezing-points, including experiments carried out at high dilutions by Bedford in 1909, gave values for i of 1·92 and thus led to the same conclusion, namely, that in solution the ions present are simply K^+ and MnO_4^-. Application of these methods by Bredig in 1893 to potassium persulphate showed that the formula of this salt must be $K_2S_2O_8$, and not KSO_4 as Marshall, who first prepared the persulphates in 1891, supposed.

In 1893 Kohlrausch showed that conductivity determinations could be applied to problems in analytical chemistry. Thus the titration of an acid by an alkali can be followed by observations on the variation of the conductivity during neutralization. As the very mobile hydrogen ions are gradually removed, the conductivity gradually sinks and ultimately attains a minimum value corresponding to the point of neutrality. Further addition of alkali causes the conductivity to rise again owing to the presence of hydroxyl ions, which are more mobile than other ions, but much less so than hydrogen ions. These methods, namely conductometric titration accompanied with observations on the change of equivalent conductivity with dilution, have been applied by Miolati and others since 1908 with the object of determining the basicity of complex acids, such as phosphomolybdic and phosphotungstic acids. It was found that when these acids are neutralized by sodium hydroxide a definite minimum value in the conductivity was observed, corresponding to the addition of 6 equivalents of alkali to one grammemolecule of the acids. Further addition of alkali caused a gradual increase in the conductivity, until 26 equivalents of alkali were added, when there was another discontinuity after which the conductivity rose rapidly with increasing quantities of alkali. It was concluded from these experiments that phosphomolybdic and phosphotungstic acids must be at least hexabasic and probably heptabasic, so that formulae such as $H_7[P(Mo_2O_7)_6]$ and $H_7[P(W_2O_7)_6]$ could be assigned to them. The function of the 26 equivalents of sodium hydroxide corresponding to the indications of the conductivity was concerned with the conversion of the complex salts into disodium phosphate and sodium molybdate and tungstate respectively.

The study of electromotive force has been inseparably connected with the early beginnings of electrochemical research. The simple forms of cells which were first used as sources of electric currents were soon found to be unsatisfactory, and hence arose the necessity of devising cells of constant electromotive force. The first satisfactory cell was devised by Daniell in 1836, and it consisted of a copper electrode immersed in a saturated solution of copper sulphate separated by a porous partition from the zinc electrode immersed in a solution of zinc sulphate. Polarization in the Daniell cell is avoided by the electrochemical replacement of zinc by copper. This cell has been the subject of numerous investigations dealing with the fundamentals of the production of electricity by chemical action, some of which are adverted to later. In 1839 Grove devised a cell in which polarization was avoided by causing the hydrogen evolved by the zinc electrode in dilute sulphuric acid to be oxidized by the action of concentrated nitric acid surrounding the platinum electrode. Other types of cells in which polarization is avoided by the use of oxidizing agents have been devised since those times. Of particular importance are cells which provide standards of constant electromotive force as distinct from producing currents. The Daniell cell can be used for approximate work in this way, but the cells devised by Latimer Clark in 1872, and by Weston in 1892 have been found to be much more reliable. The Clark cell consists of the arrangement zinc/zinc sulphate/zinc sulphate with mercurous sulphate/ mercury, and yields an electromotive force of 1·434 volts at 15° C. The Weston cell is of similar construction, but zinc and zinc sulphate are replaced throughout by cadmium and cadmium sulphate. The electromotive force of the Weston cell is 1·0185 volts at 15° C., and the variation of the electromotive force with change of temperature is considerably less than with Clark's cell. The Weston cell has accordingly displaced the Clark cell as a standard of electromotive force.

Theories concerning the action of cells began at an early stage of studies on the electric current. At the outset there were two rival theories concerning the origin of the electric energy. According to one theory the electricity was supposed to arise from contact between dissimilar metals. The other theory regarded the source of the electricity as due to chemical action. While the greater part of the progress in the study of voltaic cells has centred round the chemical theory, it should not be forgotten that contact electricity has attracted a considerable amount of attention. In 1851 Kelvin assumed that the heat of the reaction in a cell was a measure of the electromotive force. Thus in the Daniell cell, when one electrochemical equivalent of zinc replaces copper in sulphate solution, the net heat of reaction is 25,065 calories. This is the change of energy associated with the transfer of 1 faraday of electricity through the cell, and the electromotive force calculated from this is 1·085 volts in close agreement with the observed value of 1·09 volts. This conclusion reached by Kelvin is approximately but not exactly true. It originated from the principle of the conservation of energy, and would necessitate the absence of any temperature change by the cell when it was yielding a current, or in other words that the electromotive force of a reversible cell has a zero temperature coefficient. In order to view this subject in its true perspective it is necessary to appreciate the early growth of the subject of energetics.

In 1824 Carnot published his celebrated treatise entitled *Réflexions sur la Puissance motrice du Feu*, which may be described as the foundation stone of the science of thermodynamics. Ten years later Clapeyron gave a clearer expression to Carnot's reasoning by making use of graphical methods of representing the adiabatic and isothermal processes. At this time heat was regarded as a material substance, and this view must undoubtedly have been prevalent in 1840, the year in which the law of constant heat summation was enunciated by Hess. This law, which states that in a series of chemical changes the total heat which is evolved depends solely upon the initial and final states of the system, and is independent of the intermediate steps, is simply a particular case of the principle of the conservation of energy. This most important principle was developed between 1842 and 1847 as the result of the independent work of Joule, Mayer, and Helmholtz. In 1848 it was shown by W. Thomson (Lord Kelvin) that the principles of the Carnot cycle could be used to construct a thermometric scale, sometimes known as Kelvin's absolute scale of temperature, which would be wholly independent of the physical properties of substances. This scale would have a zero identical with that of a thermometer having a perfect gas as the working substance. Among the first investigators who developed correct mathematical expressions for transformations based upon the mechanical theory of heat, particular mention must be made of Clausius and of Rankine, who independently introduced the conception known as entropy, Rankine having named it the thermodynamic function. In 1849 it was shown theoretically by J. Thomson that, if a substance has a greater specific volume in the solid as compared with the liquid state, increase of pressure should lower the melting-point. This was verified by Kelvin in the following year, who showed experimentally that the application of pressure to ice resulted in the calculated degree of lowering of the melting-point. Thus arose the well-known latent heat equation $\lambda = \theta(v_2 - v_1)\dfrac{dp}{d\theta}$, in which the difference in the specific volumes of the substance in the two states v_2 and v_1 are connected with the latent heat λ, the temperature θ, and the rate of change of pressure with temperature $dp/d\theta$. This equation, which has also been attributed to Clapeyron, was developed in widely different directions by Kelvin, by Rankine, and by Clausius; and later, particularly with reference to chemical processes, by Willard Gibbs, by van't Hoff, and by Le Chatelier.

The conception of entropy was followed by that of thermodynamic potential due to Gibbs (1874–8) and by that of free energy due to Helmholtz about 1882. When Kelvin formulated his theory of the electromotive force of a reversible cell in 1851, the distinction between free energy and total energy was imperfectly understood. Gibbs and Helmholtz independently showed that it is only the *free* energy produced by the chemical changes in the cell which is used in maintaining the current, and were thus in a position to apply the necessary correction to Kelvin's reasoning. The Gibbs-Helmholtz equation, a more general theorem than the latent heat equation, can be written $E = \lambda + \theta\dfrac{dE}{d\theta}$, in which E is the electromotive force, λ the heat of the reaction, and $dE/d\theta$ the rate of change of electromotive force with temperature, and has been fully verified experimentally by several investigators, notably by Jahn in 1886.

Since a source of electromotive force can be produced by placing two metals in an electrolyte, provided that a difference of any kind exists either in the metals themselves, or in the nature or concentration of the electrolyte around them, studies on concentration cells have been of particular value in the development of the chemical aspects of electromotive force. In cells of this kind the electrical energy is derived from the energy of expansion of substances from higher to lower concentrations. A general thermodynamic theory of such cells was given by Helmholtz in 1878, and in 1889 Nernst published an important paper on the electromotive activity of the ions. He developed what he termed a theory of electrolytic solution pressure, which was a measure of the tendency of a metal to become ionized. According to Nernst the electromotive force of a concentration cell should be directly proportional to the logarithm of the ratio of the osmotic pressures of the ions in the concentrated and in the dilute solution respectively. Since the osmotic pressures were assumed to be proportional to the concentrations, the latter could be substituted for the former in the logarithmic equation. Nernst was able to bring forward a considerable amount of experimental evidence in support of his theory.

Eventually the equation assumed the form $E = \dfrac{RT}{nf} \log_e \dfrac{c}{c'}$, in which R is the gas constant expressed in electrical units, T is the absolute temperature, f is one faraday, and n is the valency of the ions. An interesting application of this principle was made by Ogg in 1898 to settle the question of the valency of mercury in the mercurous condition. From the purely chemical standpoint it appeared not unlikely that the mercurous ion was univalent, but a crucial test was obtained by studying the electromotive force of cells consisting of the combination mercury/dilute mercurous nitrate/concentrated mercurous nitrate/mercury. Ogg estimated that with a ratio of 1/10 for the concentrations of the mercurous salt in the two solutions, the electromotive force of the cell should be 0·058 volt if the mercurous ion was univalent, and half this value, viz. 0·029 volt, if the ion was bivalent. The values actually obtained were very close to the latter figure, and it was accordingly concluded that the mercurous ion must be bivalent and have the formula Hg_2^{++}.

Instead of constructing concentration cells with identical electrodes and electrolytes of different ionic concentration, another type has been realized experimentally which consists of a single electrolyte together with different concentrations of the electrodes. Thus it has long been known that if platinum or palladium containing hydrogen occluded in the metal is immersed in a liquid containing hydrogen ions, such as a dilute acid, there is a definite difference of potential between the two electrodes, so long as there is an unequal concentration of hydrogen in them. The value of the electromotive force thus produced has been shown theoretically to be proportional to the logarithm of the ratio of the pressures of the hydrogen in the two electrodes. Such a system was shown in 1893 by Le Blanc to be reversible, and has been the subject of very numerous investigations in more recent times. The hydrogen electrode has been of particular importance as a standard, the potential of which has been fixed as zero on account of the position of hydrogen in the electrochemical series of the metals. On account of the great importance of hydrogen-ion concentration in so many different branches of chemical work a standard reference electrode is a necessity. The hydrogen electrode has been associated with a number of investigators, particularly with Sörensen since 1909, whose numerous contributions to the whole subject of hydrogen-ion concentration have exerted far-reaching influences on biological science. Various other types of standard electrode have found favour with other investigators. Thus Ostwald and his pupils have made extensive use of the calomel electrode, consisting of mercury in contact with mercurous chloride in a solution of potassium chloride. In more recent times other types of reversible electrode have been found useful for particular kinds of work. Thus the quinhydrone electrode, first introduced by Biilmann in 1921, has been extensively used, particularly in studies of oxidation and reduction of organic compounds.

Of the various directions in electrochemical work in which attention to potential measurements is of particular importance two may be briefly mentioned, namely, metallic separations and potentiometric titrations. The deposition of metal on a cathode concerns the electrometallurgical industries and also the quantitative separation of metals in analysis. As early as 1883 Kiliani drew attention to the importance of adjusting the applied electromotive force to effect separations. Later, as the result of the work of Le Blanc and of Freudenberg in 1893, the theoretical foundations of this were established. Thus when a current is passed through a solution of copper and zinc sulphates, the copper is completely deposited first and the potential between the electrodes must be raised to effect the deposition of the zinc. This arises in consequence of the considerable difference between the electrode potentials of the two metals. But if the electrolysis takes place in a solution in which the metals are present as complex anions, as for instance in the presence of excess of potassium cyanide, the discharge potentials are completely altered and happen to be close together, and consequently the two metals are deposited simultaneously as an alloy. The electrodeposition of alloys has been the subject of numerous investigations, and especially interesting are some experiments carried out by Creutzfeldt in 1921, who showed that mixed crystals of variable composition can be obtained from solutions of pairs of metallic ions by variation of other conditions such as the current density and the composition of the mixtures as well as the applied electromotive force.

The first measurements of the electromotive force of oxidizing and reducing agents with a view to giving expression to the relative strengths were made by Bancroft in 1892. Although his measurements have received corrections at the hands of later investigators, the general order in which these substances were placed was substantially accurate. Potentiometric titrations were first devised by Behrend in 1893, who determined the end-point of the reaction between a chloride and silver nitrate in this way. Such methods have come into greatly extended use, particularly as the result of the work of Hildebrand and others since 1913.

MacInnes and others have shown that transport numbers can be obtained by experiments with concentration cells. Since 1915 it has become clear that calculations of electromotive force based upon the logarithmic ratio of the concentrations of the ions are by no means exact, and that it is necessary to distinguish between the *actual* and the *effective* concentration of the ions by the introduction of activity coefficients.

The application of electrolysis to preparative work extends back to the early beginnings of the subject. It is interesting to note that in numerous instances practice has preceded theory: a gifted worker has discovered some particular experimental procedure which has brought about the desired result, the theoretical explanation of which has come at a later date. In attempting to view this subject in retrospect it is desirable first to consider the types of compounds which can be electrolyzed. Broadly speaking these consist of solutions, particularly aqueous solutions, of acids, bases, and salts, and also of fused salts. The conduction of electricity by such substances, as distinct from conduction by metals, is always accompanied with chemical change. The question as to whether there is a sharp distinction between metallic and electrolytic conduction, or whether there may exist some types of compounds in which some gradual transition from purely metallic to purely electrolytic conduction is to be observed, occupied the attention of Faraday about 1833 and of Hittorf some twenty years later, both of whom considered that such a transition might be found among compounds of the type of the metallic sulphides and selenides. The experimental evidence was, however, by no means convincing. Nernst in 1898 considered that the mixture of metallic oxides used in his lamp conducted electricity like an electrolyte. For many years it was accepted that the conduction of electricity by alloys and amalgams was wholly metallic and non-electrolytic, and the conclusions were perfectly correct as far as they went. In 1924, Kremann and his collaborators carried out a series of experiments on the passage of heavy currents through various alloys, and obtained conclusive evidence of electrolysis. Thus with antimony-zinc alloys it was found that the zinc moved towards the cathode and the antimony towards the anode, thus enabling a partial separation to be effected. With other alloys and with amalgams the direction of movement of the metals was such as might be expected having regard to their electrochemical character. Thus with sodium amalgam the sodium content is increased at the cathode. These experiments will bear comparison with some investigations carried out by Moers in 1920 on the hydrides of the alkali metals. He found that when lithium hydride was electrolyzed, the hydrogen migrated towards the anode and the lithium towards the cathode. In this case hydrogen functions as the *electronegative* constituent of the compound, and quantitative experiments carried out by Peters in 1924 showed that the electrolysis followed the requirements of Faraday's law as measured by the rate of evolution of the gas.

Some of the earliest experiments of a preparative character appear to have originated from observations on the effects produced by anodic oxidation and cathodic reduction. Thus Kolbe in 1848, using Bunsen cells as his source of current, observed the formation of potassium chlorate during the electrolysis of solutions of potassium chloride. He admitted that the production of this compound might have been due to a secondary reaction between chlorine and potassium hydroxide, but this objection did not apply to other experiments carried out in which he obtained potassium perchlorate. In 1848–50, Kolbe carried out numerous experiments on the electrolysis of potassium acetate and valerate, the object of which was to study the effects of oxidizing the acids electrolytically. He obtained a gas which he termed methyl from acetic acid and a high boiling liquid which he named valyl from 'valerianic' acid. It is obvious from his description that Kolbe used isovaleric acid,

$$(CH_3)_2CH.CH_2.CO_2H,$$

and the compound which he termed valyl and formulated according to his system as C_8H_9 was the octane, di-isobutyl,

$$(CH_3)_2CH.CH_2.CH_2.CH(CH_3)_2.$$

The experiments were carried out in cells divided by porous partitions to enable him to ascertain the nature of the anode and cathode reactions respectively. He found that hydrogen was evolved at the cathode and a mixture of carbon dioxide and 'valyl' at the anode; but he was very careful to specify the conditions necessary for securing this result, and remarked that the electrolysis might proceed in a very different manner with altered experimental conditions. In the case of potassium acetate he observed the liberation of hydrogen at the cathode and a mixture of carbon dioxide and 'methyl', which he formulated as C_2H_3, at the anode. Having regard to the date at which these experiments were made, and the condition of chaos regarding chemical formulae at that time, these results are of exceptional interest. In the light of modern formulae the production of ethane by the union of the methyl groups from two molecules of acetic acid, and a similar doubling of two isovaleric residues to produce di-isobutyl, is of much importance, and must have inspired Crum Brown and J. Walker to investigate the electrolysis of the potassium and sodium alkyl esters of the dibasic acids, which they began in 1890. Thus they obtained a 60 per cent yield of diethyl succinate from potassium ethyl malonate, and a 35 per cent yield of diethyl adipate from potassium ethyl succinate. Here, as always, attention to the experimental conditions was found to be very necessary. In 1893, Walker extended this work to investigate the nature of camphoric acid, the basicity of which was uncertain at that time. It was considered that if camphoric acid were dibasic, it should resemble other dibasic carboxylic acids when its sodium ethyl ester was submitted to electrolysis. This was found to be correct by Walker and his collaborators between 1895 and 1900, although some of the results were difficult to interpret. In the electrolysis of dibasic acids according to the procedure of Crum Brown and Walker, the formation of unsaturated acid esters as well as esters derived from acids having double the number of methylene groups in the molecule had been observed. Camphoric acid, $C_8H_{14}(CO_2H)_2$, was found to be peculiar in giving rise to isomeric sodium ethyl esters, which were known as the *ortho* and *allo* compounds respectively. The products obtained by Walker by electrolyzing solutions of *ortho*-sodium ethyl camphorate consisted chiefly of the ethyl esters of a monobasic unsaturated acid, $C_8H_{13}CO_2H$, and of a dibasic acid, $C_{16}H_{28}(CO_2H)_2$, which were termed campholytic acid and camphothetic acid respectively. The electrolysis of the *allo*-ethyl potassium camphorate took a similar course, and yielded an acid isomeric with the monobasic campholytic acid obtained from the *ortho* compound. The acid derived from the *ortho*-sodium ethyl salt was found to be optically inactive, and also according to Walker to exist in two isomeric forms, namely as ordinary campholytic acid and as *iso*-lauronolic acid. Some of these results, particularly the problem of *iso*-lauronolic acid, were regarded as of much importance at that time in connexion with the establishment of the constitutional formula of camphor.

The mechanism of the Kolbe electrosynthesis has been discussed in terms of two altogether different theories, which for brevity may be named as the oxidation and the discharged anion theories respectively. It will be remembered that Kolbe began his experiments with the object of submitting the carboxylic acids to anodic oxidation. Crum Brown and Walker were definitely of opinion that the formation of hydrocarbons and esters was due to direct union between the discharged anions, the separation of carbon dioxide taking place simultaneously. For many years since that time advocates of these rival theories have been forthcoming. In 1892, Murray investigated the best conditions for preparing ethane by electrosynthesis, and he observed that when a solution of acetic acid containing free sulphuric acid was electrolyzed no appreciable oxidation of the acetic acid took place. It is well known that the fatty acids are highly resistant to attack by oxidizing agents. Murray's results have been quoted by later investigators as evidence against the oxidation theory. In 1918, Fichter and Krummenacher put forward an oxidation theory involving the mechanism of per-acid formation, which recalls the recognition of acidic peroxides by Brodie some fifty years previously. In 1925 and 1926, investigations were carried out by Gibson and by Robertson and also by Fairweather and O. J. Walker on the mechanism of the Kolbe process. They took careful account of the various factors which influence any electrolytic process, namely, the nature of the electrodes, the current density, the difference of potential between the electrodes, and the concentration and temperature of the electrolyte. These investigations were carried out under the direction of J. Walker, and opposite conclusions were reached. Gibson and Robertson considered that the process involved oxidation, whereas Fairweather and O. J. Walker were convinced that the mechanism consisted essentially in the union of the discharged ions.

Although the persulphates were first prepared and isolated by electrolysis of acid sulphate by Marshall in 1891, the beginnings of this work are to be traced to very much earlier dates. It was known to Faraday that if sulphuric acid of a fair degree of concentration was electrolyzed, the proportions by volume of hydrogen and oxygen liberated from the cathode and anode respectively were no longer in the familiar ratio of 2/1. Thus he remarked that 'if the acid were very strong, then a remarkable disappearance of oxygen took place; thus, one made by mixing two measures of strong oil of vitriol with one of water, gave forty-two volumes of hydrogen, but only twelve of oxygen...'. In 1878, Berthelot recognized that the disappearance of some of the oxygen which occurs during the electrolysis of moderately concentrated sulphuric acid was not due to the formation of hydrogen peroxide, as had been supposed by some, but to the formation of what he termed *acide persulphurique*. He imagined that the acid might have the formula HSO_4 or $H_2S_2O_8$, and he also recognized that a second oxidation product was also produced, which differed from the first in having more powerfully oxidizing properties. Between 1889 and 1893, Traube investigated the products present in sulphuric acid which had been submitted to electrolysis, and obtained somewhat conflicting results for the ratio of sulphate to active oxygen. In the meantime Marshall had prepared the alkali persulphates in crystalline condition, observing the necessity of high current density at the anode and a low temperature to obtain satisfactory results. Later, especially in 1898, as a result of the work of Caro it was found that there are two persulphuric acids, namely, the acid prepared by Marshall now usually known as perdisulphuric acid, and permonosulphuric acid, H_2SO_5, frequently termed Caro's acid, which is derived from the former by hydrolysis in concentrated acid solution. The mechanism of the formation of the persulphates by electrolysis has presented a problem not unlike that of the mechanism of the Kolbe electrosynthesis. In solutions of bisulphates, particularly with a fairly high concentration of sulphuric acid, the ion HSO_4^- is undoubtedly present. The formation of a persulphate could therefore readily take place by the union of two discharged anions, and for some time the production of potassium persulphate was supposed to proceed in this way. Later opinion seemed to favour a mechanism based upon the action being due to direct anodic oxidation. In somewhat indirect support of this view it may be noted that Fichter and Humpert in 1923 showed that potassium bisulphate can be oxidized to the persulphate directly by passing fluorine into a concentrated solution of the salt.

The question as to whether preparative work is better carried out by purely chemical as distinct from electrochemical methods has frequently been discussed; and as far as any general answer can be given to such a problem, it would seem to be correct to say that when the various factors which determine favourable conditions for carrying out some particular reaction have been worked out, the electrochemical method is to be preferred, because the products can then be obtained in purer condition. A good example of this is to be found in Haber's scheme for the electrolytic reduction of aromatic nitro-compounds, which was worked out in 1898. From nitrobenzene it is possible to obtain nitrosobenzene, phenylhydroxylamine and its isomeride *p*-aminophenol, azoxybenzene, azobenzene, hydrazobenzene and the isomeric benzidine, and ultimately aniline. As regards purely chemical methods of reduction, it can be stated generally that powerful reducing agents, particularly in acid solution, will effect complete reduction to the amino compound, whereas milder reducing agents, especially in nearly neutral or in alkaline solution, will give rise to the intermediate products. The electrolytic reduction of aromatic nitro-compounds has been the subject of many investigations, but the essential features of the reactions were clearly shown by Haber, who was able to distinguish between products which are formed by direct reduction and those which resulted from secondary chemical changes. The direct products, namely, nitrosobenzene, phenylhydroxylamine, and aniline, all contain a single benzenoid nucleus; those with two phenyl groups in the molecule, viz. azoxybenzene and azobenzene, are formed by condensation between two or more molecular proportions of the direct products. Azoxybenzene is readily reduced electrolytically to hydrazobenzene. In strongly acid solution phenylhydroxylamine is transformed into *p*-aminophenol, and a similar transformation of hydrazobenzene into benzidine also takes place.

In the preparation of many of the elements, particularly the metals, electrolytic methods have been of outstanding importance. From the isolation of the alkali metals by Davy in 1807 to the preparation of radium by the electrolysis of a solution of the chloride, using a mercury cathode, by Madame Curie and Debierne in 1910—just over a century—many elements hitherto only recognized in their compounds have been isolated by electrolysis. Thus although aluminium had been isolated by the action of potassium on the chloride by Wöhler in 1828, Bunsen prepared the metal by the electrolysis of fused sodium aluminium chloride in 1854. This double salt is not suitable for the preparation of the metal on a technical scale. Some thirty years later the electrolytic preparation of aluminium became a commercial success by using an electrolyte consisting of bauxite (aluminium oxide) dissolved in molten cryolite (sodium aluminium fluoride). In 1861, Crookes obtained thallium by the electrolysis of solutions of thallous sulphate which he prepared from residues of flue dust from sulphuric acid works where pyrites containing the metal had been burnt. The isolation of fluorine—an operation of considerable experimental difficulty— was accomplished by Moissan in 1886 after numerous unsuccessful attempts had been made by other investigators. This very reactive gas was prepared by the electrolysis of a solution of potassium fluoride in anhydrous hydrogen fluoride kept at a low temperature by the aid of a bath of boiling methyl chloride. The electrolyte was placed in a copper vessel and platinum electrodes were used.

In reviewing the history of electrochemical research it should never be forgotten that most of the work in the nineteenth century was accomplished with primary batteries as the source of current. Lead accumulators, so important and familiar in modern times, were a relatively late development. The first experiments which resulted in the invention of these secondary cells were made by Planté in 1860, but it was not until some twenty years later that the improvements devised by Faure resulted in the lead accumulator becoming a reliable source of constant current. The theory of the chemical changes which take place during the charge and discharge of lead cells was first put forward by Gladstone and Tribe in 1883. This theory gives expression to the formation of lead dioxide by anodic oxidation together with the liberation of free sulphuric acid during the charging process, and the removal of these substances during discharge in terms of the reversible reaction:

$$2PbSO_4 + 2H_2O \underset{\leftarrow \text{discharge}}{\overset{\rightarrow \text{charge}}{\rightleftarrows}} PbO_2 + Pb + 2H_2SO_4.$$

This equation undoubtedly represents the chief reactions during charge and discharge, but there are certain difficulties, such as the sparing solubility of lead sulphate and the exceptionally high electromotive force of freshly charged cells, which have received attention at the hands of later experimenters. Considered thermodynamically the lead accumulator approximates sufficiently to a reversible system to justify the application of the Gibbs-Helmholtz equation to the problem. In 1898–9 Dolezalek studied lead cells from the standpoint of energetics, and obtained satisfactory agreement between the values of the electromotive force as determined experimentally and as calculated from the concentrations of the sulphuric acid. The general correctness of the elementary theory of the lead accumulator thus received confirmation, but a certain amount of dissatisfaction still remained. Suggestions regarding the active material in the positive electrodes to be some oxide of lead other than the dioxide, possibly Pb_2O_5, have been made, but have not received any very convincing experimental verification. Glasstone in 1922, however, brought forward some evidence in favour of the view that the initial high electromotive force of a freshly charged accumulator may be due to the presence of an extremely small quantity of some higher oxide in solid solution in the dioxide.

The application of electricity to chemical industry and the production of electricity by chemical action have in many ways been interwoven in the past, but in modern times the importance of the former has been in the ascendant. Faraday's experiments on electromagnetic induction, begun in 1831, gave rise to the dynamo, and this together with the invention of the lead accumulator have completely displaced primary batteries as a source of current on any considerable scale. Nevertheless, dry cells, consisting of zinc and carbon electrodes, the former being immersed in an electrolyte in a paste or jelly and the latter being surrounded with manganese dioxide to act as a depolarizer, have acquired a greatly extended use in recent years, particularly for use in flash lamps, in wireless receiving sets, and in other applications where currents are required intermittently. As regards the production of electricity on a large scale the problem is that of the conversion of mechanical energy into electrical energy. The two chief sources of mechanical energy are coal and water power. In countries which have no coal but are well endowed with water power the installation of hydro-electric plant has soon taken place. Where both coal and water power are available opinion has differed regarding the relative advantages of generating electricity for chemical or metallurgical industries by steam raising or by the direct use of hydraulic power. The mechanical energy of even a good steam engine represents at best a very small fraction of the thermal energy derived from the combustion of the coal. It is true that from this point of view internal combustion engines compare favourably with those driven by steam, but are disappointing none the less. This has given rise to what has been termed the problem of the fuel cell. If the oxidation of carbon or of carbon monoxide to the dioxide could be made to take place by an electrolytic process, instead of by combustion, it has been urged that this considerable wastage of thermal energy would be obviated, and the electricity would be obtained very much more cheaply. Many investigators have attacked this problem, and so far without success. One single example may be quoted by way of illustration. Hofmann in 1918 showed that carbon monoxide is oxidized to the dioxide by air in presence of alkali at a copper surface. He constructed a cell consisting of two copper electrodes dipping in an alkaline liquid, one being surrounded by air and the other by carbon monoxide. At room temperature this cell gave rise to an electromotive force of 1·32 volts as compared with a value of 1·34 volts calculated by Nernst and Wartenberg. The cell was of no practical value, however, on account of its high resistance and the slowness of the oxidation. In 1922, Rideal and Evans directed attention to the problem of the fuel cell, and analysed the probable reasons for the failure of such cells to become a practically useful source of current.

REFERENCES

M. FARADAY. *Experimental Researches in Electricity*. Republished by Messrs J. M. Dent and Sons Ltd. in Everyman's Library.

E. T. WHITTAKER. *A History of the Theories of Aether and Electricity*. Dublin, 1910.

J. LARMOR. *Aether and Matter*. Cambridge, 1900.

W. C. D. WHETHAM. *A Treatise on the Theory of Solution including the Phenomena of Electrolysis*. Cambridge, 1902.

F. FOERSTER. *Elektrochemie Wässeriger Lösungen*, Dritte Auflage. Leipzig, 1922.

A. J. ALLMAND and H. J. T. ELLINGHAM. *The Principles of Applied Electrochemistry*. London, 1924.

P. WALDEN. *Salts, Acids, and Bases*. New York, 1929.

H. FALKENHAGEN. *Electrolytes*. Translated by R. P. Bell. Oxford, 1934.

S. GLASSTONE and A. HICKLING. *Electrolytic Oxidation and Reduction. Inorganic and Organic*. London, 1935.

H. KOLBE. *The Electrolysis of Organic Compounds*. Alembic Club Reprints, No. 15.

F. Kohlrausch und L. Holborn. *Das Leitvermögen der Elektrolyte,* Zweite Auflage. Leipzig, 1916.

S. Glasstone. *The Electrochemistry of Solutions.* Second Edition, London, 1937.

J. H. van't Hoff and Svante Arrhenius. *The Foundations of the Theory of Dilute Solutions.* Alembic Club Reprints, No. 19.

G. F. FitzGerald. Helmholtz Memorial Lecture. *J. Chem. Soc.* 1896, p. 885.

S. Arrhenius. The Faraday Lecture. *J. Chem. Soc.* 1914, p. 1414.

Annual Reports of the Chemical Society since 1904. Especially a report by D. M. Murray-Rust, O. Gatty, W. A. Macfarlane, and H. Hartley on The Electrical Conductivity of Solutions, 1930.

Chapter III

STEREOCHEMISTRY

The beginnings of stereochemistry are to be found in polarimetric experiments which were started by Biot about the year 1815. This physicist found that certain substances can cause rotation of plane polarized light. Such substances were found to be of two classes, namely, those which cause rotation only in the crystalline condition, and others which possess the property under any conditions. To the latter class belong substances such as sugar, camphor and tartaric acid. This acid had been studied about 1844 by Mitscherlich as regards its optical and crystallographic properties, who recognized that racemic acid (then known as paratartaric acid) was identical with ordinary tartaric acid in nearly all respects, except as regards its behaviour towards polarized light. This subject attracted the attention of Pasteur who made a very exhaustive study of various tartrates, especially the double sodium ammonium salt. As the result of very careful work between 1848 and 1853, Pasteur correlated certain peculiarities of the crystals with differences in optical properties, and in particular he showed that when a solution of sodium ammonium racemate was allowed to crystallize at the ordinary temperature, the saturated solution deposited separate crystals of *dextro*-sodium ammonium tartrate and of *laevo*-sodium ammonium tartrate. He also elaborated two other methods for effecting the resolution of racemic acid, namely, by the use of organisms which behave differently with the two optically active constituents, and by crystallization with optically active bases, such as cinchonine or strychnine. Pasteur in 1853 discovered a non-resolvable modification of tartaric acid (mesotartaric acid) which he obtained by heating ordinary tartaric acid with water for some time. In 1860, Pasteur made a considerable approach towards a satisfactory theory of these remarkable phenomena, and recognized the existence of molecular dissymmetry. He attached great importance to the identity of the crystalline properties of the dextro and laevo acids, the sole difference between them consisting in the right- and left-handed disposition of their hemihedral faces corresponding with their right- and left-handed optical rotatory properties, and even suggested a comparison of the arrangement of the atoms in the molecule of *dextro*-tartaric acid with that of a right-handed spiral.

More than one cause is to be ascribed to theoretical developments subsequent to the fundamental work of Pasteur. Lactic acid derived from sour milk was known to be identical with lactic acid obtained from extract of meat, except that the former was optically inactive and the latter was dextro-rotatory. Wislicenus in 1873 recognized that plane constitutional formulae were inadequate to discuss isomerism of this kind, and pointed out that spatial considerations were a necessity. There can be no doubt that Kekulé, whose celebrated formula for benzene appeared in 1865, had been giving much thought to the disposition of the valency bonds of the carbon atom, and in 1867 had even given expression to the idea of some sort of tridimensional, possibly tetrahedral, configuration. The inadequacy of a planar arrangement for the valencies of the carbon atom in the molecule of methane was also recognized about 1871 by Victor Meyer, who realized that such an arrangement would involve the existence of two isomeric methylene chlorides. Such was the position when the whole field was clarified in 1874 independently by Le Bel and van't Hoff.

Both van't Hoff and Le Bel were working in Wurtz's laboratory in Paris almost up to the date of the separate and nearly simultaneous publication of their theoretical ideas. It is very remarkable that the two men were on intimate terms with one another, and yet each was wholly ignorant of any interests of the other in this department of chemistry. Although the ideas of the two chemists were fundamentally identical, there were shades of difference in the manner of their approach. Van't Hoff started his ideas from Kekulé's doctrine of the quadrivalency of carbon, and particularly from Kekulé's added hypothesis of 1867 regarding the direction of the valencies towards the corners of a tetrahedron. He was doubtless also influenced by the views of Wislicenus regarding the isomerism of the lactic acids. Le Bel was more especially influenced by the researches of Pasteur, and his whole treatment of the subject was more purely geometrical than that adopted by van't Hoff. The outstanding contribution of both men was to define the conditions under which optical isomerism can arise, namely, when the molecule of the compound contains one or more asymmetric carbon atoms, i.e. atoms having four univalent atoms or groups, all of which are different, united to each of the four valencies of the carbon atoms. In detail it may be noted that whereas we are indebted to van't Hoff for explaining the nature of the isomerism of unsaturated compounds, such as maleic and fumaric acids, and thus laying the foundations for the study of geometrical as distinct from optical isomerism, to Le Bel must be given the credit for explaining the nature of optical inactivity due to internal compensation, such as that of mesotartaric acid, as distinct from the inactivity of racemic acid which is a consequence of external compensation (see p. 70).

From these beginnings in 1874, the study of optical isomerism in which the relations between dextro- and laevo-isomerides are as those of object and mirror image, and of geometrical isomerism in which optical activity does not necessarily arise, but in which there are nevertheless differences in properties, has taken widely different directions. It will be interesting to follow some of these, but it should be remembered that although the ramifications are considerable, yet there is a fairly close relationship between many of them. The theory of the asymmetric carbon atom had at first anything but a favourable reception. But it is interesting to recall that the strictures of Kolbe, who poured ridicule upon it with as much vigour as he had previously done upon certain structural formulae, possibly attracted the attention of some who might have otherwise neglected the new theoretical ideas. It was, however, eagerly welcomed by Wislicenus, who between 1887 and 1889 devoted much attention to the study of geometrical isomerides of the type of maleic and fumaric acids, and pointed out that maleic acid must have the *cis* configuration in which the two carboxyl groups are on the same side of the double bond because of the readiness with which the acid forms an anhydride, whereas fumaric acid must have the *trans* configuration with the carboxyl groups on opposite sides of the double bond.

On the experimental side, it may be said that the original methods of resolution devised by Pasteur, with some minor modifications and extensions, have been responsible for the erection of the vast edifice of stereochemistry—a term due to Victor Meyer—which exists at the present time. The original method of crystallization has very limited application, but in a modified way it was used with success by Purdie in 1893 in resolving ordinary lactic acid into its active components by introducing a nucleus of one of them into a supersaturated solution of the double zinc ammonium salt. The method of fractional crystallization of salts formed between the components of an inactive mixture or compound of acids with optically active bases, or similarly with the salts derived from the components of inactive products of bases with optically active acids, has received a great deal of attention. It has also been developed in other directions. Thus between 1899 and 1901 Marckwald and McKenzie observed that the rate of esterification of the dextro- and laevo-constituents of an inactive mixture of alcohols with an optically active acid was different, and they applied this successfully to effect resolution.

Many naturally occurring organic substances are optically active, whereas when compounds having molecular dissymmetry are formed synthetically equivalent quantities of the dextro- and laevo-isomerides are invariably produced. In what respects do the conditions of formation of natural products differ from the conditions of the laboratory? This problem has engaged the attention of chemists since the time of Pasteur, who indeed realized that in Nature the production of substances such as sugars and alkaloids must take place under some sort of asymmetric conditions, and it is said that he actually carried out some experiments of a preparative character in powerful magnetic fields in order to imitate such conditions. In 1894, Fischer introduced a conception of asymmetric synthesis in one of the senses in which this term is understood at the present time, namely, the production of optically active substances without the necessity of employing one of Pasteur's methods of resolution. Fischer, in considering the synthesis of sugars in the plant, regarded the process as taking place through the agency of the reduction of atmospheric carbon dioxide to formaldehyde, followed by the polymerization of this to carbohydrates. This latter process was regarded by him as subject to the directing influence of the chlorophyll already present in the cells, and in such a manner that each successive asymmetric carbon atom in a chain being produced with an excess of one of its optical iso-merides over the other. The net effect is therefore the production of an optically active sugar, the chlorophyll being regenerated in the process.

After a number of unsuccessful attempts on the part of several investigators, a result of great interest was obtained by Marck-wald in 1904. On heating the acid brucine salt of the potentially inactive methyl-ethyl-malonic acid,

$$\begin{array}{c} CH_3 \\ \diagdown \\ C \diagdown \\ C_2H_5 \diagup \diagdown CO_2H \\ CO_2H \end{array},$$

until the evolution of carbon dioxide was complete, and liberating the valeric (methyl-ethyl-acetic) acid thus produced, the latter was found to contain a considerable excess of the laevo-isomeride. This result was subjected to some rather stupid criticism at the time, but it is now generally recognized that Marckwald's result may rightly be termed an asymmetric synthesis. In this con-nexion it may be relevant to mention Marckwald's definition of an asymmetric synthesis which has since been frequently quoted: 'Asymmetrische Synthesen sind solche, welche aus symmetrisch konstituirten Verbindungen unter intermediärer Benutzung optisch aktiver Stoffe, aber unter Vermeidung jedes analytisches Vorganges, optisch-aktive Substanzen er-zeugen.'

Since 1904 numerous asymmetric syntheses were effected by McKenzie and his collaborators. Thus in 1906, McKenzie and Wren in studying the reduction of the optically active esters of certain α-ketonic acids found that when the borneol ester of pyruvic acid was reduced, the resulting lactic acid ester acetyl-ated and then hydrolyzed, an optically active lactic acid was obtained.

It would seem that the formation of optically active compounds in Nature is frequently capable of being explained on the basis of inequality in the velocity of reaction of synthesis (or of destruction). In this connexion some experiments carried out in 1908 by Bredig and Fajans on the decomposition of camphor-carboxylic acid into camphor and carbon dioxide,

$$C_8H_{14} \begin{array}{c} CH.CO_2H \\ | \\ CO \end{array} \rightarrow C_8H_{14} \begin{array}{c} CH_2 \\ | \\ CO \end{array} + CO_2,$$

are of great interest. This reaction is accelerated by bases, and it was found that if optically active bases, such as nicotine or quinine, were used, the catalytic effect was different with the optical isomerides of the camphorcarboxylic acid. Thus they found that when inactive camphorcarboxylic acid was decomposed in the presence of quinine, the camphor formed was slightly laevo-rotatory and the remaining acid slightly dextro-rotatory, and they considered that this catalytic decomposition was closely analogous to an enzyme reaction.

All the examples which have been so far considered have sometimes been classified under the heading of *partial* asymmetric synthesis because they have been effected under the influence of previously existing optically active compounds including enzymes, in contradistinction to *total* or *absolute* asymmetric synthesis in which no previously existing optically active compound has been introduced at *any* stage. In 1894, van't Hoff ventured the opinion that the origin of optically active substances in Nature might be traced to the action of circularly polarized light. In 1896, Cotton discovered the effect which is known by his name, viz. that in the neighbourhood of certain absorption bands of optically active compounds circular dichroism for light of some particular wave-length will be exhibited. It will be seen in what follows that after numerous unsuccessful experiments, which it is not necessary to describe, undoubted examples of absolute asymmetric synthesis were realized nearly forty years after the discovery of the Cotton effect. An important step was taken by Byk in 1904, who pointed out that since sunlight reflected by the surface of the sea is partially plane polarized the effect of terrestrial magnetism is to cause this reflected light to become partially elliptically polarized, and further that dextro-circularly polarized light must predominate at the surface of the earth. Byk considered that this unsymmetrical photochemical energy might be utilized in the synthesis of optically active compounds in organisms.

Of the numerous unsuccessful experiments to realize an absolute asymmetric synthesis, it was pointed out by Byk in 1909 that negative results must follow from reactions carried out in *ordinary* light in a magnetic field, since such a combination will not give rise to circularly polarized light. The first successful result was obtained by Kuhn and Braun in 1929 who found that when racemic ethyl α-bromopropionate was illuminated by circularly polarized light a very feebly active product was produced. This may be termed an asymmetrical photochemical decomposition, but on account of the minuteness of the effect observed the result was not very convincing. In 1930, Kuhn and Knopf obtained a much more definite result with α-azido-propionicdimethylamide, $CH_3 . CHN_3 . CO . N(CH_3)_2$, in an inert solvent. When irradiated with dextro- or with laevo-circularly polarized light of wave-length corresponding to that of the absorption band so as to effect about 40 per cent decomposition of the compound, it was found that the sign and value of the rotatory power of the undecomposed amide were of the expected order of magnitude.

Dealing with the origin of naturally occurring optically active compounds in his Presidential Address to the Chemistry Section of the British Association in 1932, Mills expressed doubts regarding the efficacy of circularly polarized light, particularly on account of the 'minuteness of the proportion of the total illumination received by an organism under natural conditions that can be circularly polarized'. He ventured the opinion that the optical activity of vital products is an inevitable consequence of the property of growth possessed by all living matter.

In a very interesting monograph on asymmetric synthesis published by Ritchie in 1933, the author remarked that in view of the positive results now realized in absolute asymmetric synthesis, it is no longer necessary to regard the synthesis of sugars in the plant as depending upon the directing effect of the chlorophyll granules, as Fischer did in 1894, to account for their optical activity. After reviewing the subject generally, Ritchie expressed the opinion that explanations based on the action of circularly polarized light are more convincing than the considerations put forward by Mills.

A problem of long standing, namely, the simplest type of compound which should be capable of displaying optical activity, received a considerable amount of attention before it was successfully solved. For many years the simplest known optically active substance was lactic acid. The molecule of this compound contains three carbon atoms, viz. two in addition to the asymmetric one, and there was a fairly widespread idea that optical activity could not be realized with a simpler molecular constitution than this. In 1914, however, it was shown by Pope and Read that when a single carbon atom is united to three different elementary atoms and to an inorganic radical, the resulting molecule can exist in active forms. The compound with which this important result was realized was chloroiodomethane-sulphonic acid, $\begin{smallmatrix} Cl \\ I \end{smallmatrix} \!\!>\!\! C \!<\!\! \begin{smallmatrix} H \\ SO_3H \end{smallmatrix}$.

In the early part of the present century, when a considerable amount of information regarding optically active substances had been accumulated, it was gradually realized that the cause of optical activity in a molecule is not to be located in the presence of one or more asymmetric atoms, but rather to the enantiomorphous configuration of the molecule considered as a whole. Indeed, such an outlook was indicated by van't Hoff in developing his tetrahedral ideas. It is now recognized that optical isomerism can arise in any molecule which is devoid of symmetry of any kind. This is admirably illustrated with reference to derivatives of allene, $H_2C:C:CH_2$, and also with spiranes.

Van't Hoff pointed out that in an allene of the type

$$\begin{smallmatrix} A \\ B \end{smallmatrix} \!\!>\!\! C \!\!=\!\! C \!\!=\!\! C \!<\!\! \begin{smallmatrix} X \\ Y \end{smallmatrix},$$

if the groups A and B are in the plane of the paper, the groups X and Y must be in a plane at right angles to that of the paper. The molecule would have no plane of symmetry, and optical activity should therefore arise. Experimental difficulties delayed the realization of the synthesis and resolution of this type of compound for many years. In 1909, however, Perkin, Pope, and Wallach succeeded in preparing 1-methyl-*cyclo*hexylidene-4-acetic acid,

$$\begin{array}{c} CH_3 \\ H \end{array}\!\!>\!\!C\!\!<\!\!\begin{array}{c} CH_2.CH_2 \\ CH_2.CH_2 \end{array}\!\!>\!\!C\!\!=\!\!C\!\!<\!\!\begin{array}{c} H \\ CO_2H \end{array},$$
$$\quad(1)\qquad\qquad(4)\ (7)$$

and resolving it into its optical antipodes. From the constitution of this compound it will be evident that if the bonds of the hexamethylene ring lie in the plane of the paper, the methyl group and hydrogen atom attached to carbon atom (1) will be at right angles to this plane, and the hydrogen atom and carboxyl group attached to the terminal doubly linked carbon atom (7) will be in the plane of the paper. This is therefore an example of an elaborated allene type of compound. A still more interesting example, a true asymmetric allene, was synthesized and resolved by Maitland and Mills in 1935, namely, the compound diphenyldinaphthylallene,

$$\begin{array}{c} C_6H_5 \\ C_{10}H_7 \end{array}\!\!>\!\!C\!\!=\!\!C\!\!=\!\!C\!\!<\!\!\begin{array}{c} C_6H_5 \\ C_{10}H_7 \end{array}.$$

An interesting example of the spirane type of compound, which resembles a figure of eight with the carbon atom at the centre having one of the loops in a plane at right angles to that of the other, is to be found in the keto-dilactone of benzophenone-2:4:2′:4′-tetracarboxylic acid,

$$HO_2C\!-\!\!\bigcirc\!\!\overset{CO-O}{\underset{O-CO}{\diagdown C\diagup}}\!\!\bigcirc\!\!-CO_2H,$$

which was resolved into its optical antipodes by Mills and Nodder in 1920. Another noteworthy example of spiro-asymmetry in which optical activity was realized in 1931 by Pope and Whitworth is *spiro*-5:5-dihydantoin,

$$\begin{array}{c} NH-CO \\ CO-NH \end{array}\!\!>\!\!C\!\!<\!\!\begin{array}{c} NH-CO \\ CO-NH \end{array}.$$

While the attention of chemists was still focused upon asymmetric carbon atoms as the source of optical activity rather than upon the wider conception of molecular dissymmetry, it was natural to extend the original conception of asymmetric atoms to those of elements other than carbon. Successful results were realized with compounds of several of such elements, the first of which to be studied were those of nitrogen. Both tervalent and quinquevalent nitrogen compounds have presented stereochemical problems, and it will be seen in what follows that an immense amount of experimental work has been centred round the problem as to the planar or spatial disposition of the three valencies of tervalent nitrogen, a question which had engaged the attention of van't Hoff, and it is interesting to recall that the first attempt to resolve a quinquevalent nitrogen compound was made in 1891 by the other founder of the tetrahedral conception, Le Bel.

When a tertiary amine in which all three substituents are different, viz. Nabc, is treated with an alkyl halide dX a quaternary ammonium salt N$abcdX$ is obtained. It goes without saying that in a molecule of this kind there are obvious possibilities of optical isomerism, but it is less simple to determine the geometric configurations. In the *Ansichten über die organische Chemie* (1878), van't Hoff attempted to discuss the stereochemistry of quinquevalent nitrogen by imagining the atom of nitrogen within a cube with the five valencies directed to five of the corners of the cube. Another suggestion due to Willgerodt in 1890 was to use a double tetrahedron, in which the nitrogen atom is in the plane between the superposed tetrahedra, and the three original valencies of the nitrogen atom directed towards the angles of the triangle thus formed, the fourth and fifth valencies being directed towards the vertices of the tetrahedra, these particular valencies thus being in a direction at right angles to the plane between the two tetrahedra. A third arrangement was suggested in the same year by Bischoff consisting of the nitrogen atom within a square pyramid, the halogen being placed at the vertex of the pyramid, and the four univalent groups at the four angles of the square base. The pyramidal configuration has certain obvious advantages over the cubical and double tetrahedral configurations as being the most symmetrical, and affording the smallest possible number of isomers, and it was used for a number of years with considerable success in the development of the subject, although, as will be seen later, investigations by Mills and Warren in 1925 led to its abandonment.

In 1891, Le Bel announced that he had obtained optically active methylethylpropylisobutyl ammonium salts. The resolution was effected by causing *Penicillium glaucum* to grow in solutions of the chloride of the base and a feeble laevo-rotation was obtained. This result was challenged by others, and it now seems doubtful if Le Bel's work was accurate. The results obtained by Pope and his pupils between 1899 and 1901 were of a much more decisive character. About the same time Wedekind had prepared benzylallylmethylphenyl ammonium bromide, but he failed to resolve it. The successful results obtained by Pope and his pupils were a consequence of the use of strong acids, namely, camphorsulphonic and bromocamphorsulphonic acids, for effecting resolution. The iodide of Wedekind's base, dissolved in a non-aqueous solvent, was treated with the silver salt of *dextro*-camphorsulphonic acid. On fractionally crystallizing the *dextro*-camphorsulphonates of the quaternary base thus produced and then removing the acid by adding potassium iodide, the optically active benzylallylmethylphenyl ammonium iodides were obtained. From 1903 onwards a very exhaustive study of the optical activity of quinquevalent nitrogen compounds was carried out by Jones. He soon found that the Bischoff pyramidal configuration for the asymmetric nitrogen atom, adopted by him in 1905, provided a fairly satisfactory basis for most of the examples of optical isomerism so far studied. A new type of optically active nitrogen compound, exemplified by methylethylaniline oxide, $\begin{smallmatrix} CH_3 \\ C_2H_5 \\ C_6H_5 \end{smallmatrix}\!\!>\!\!N\!\!=\!\!O$ was, however, resolved by Meisenheimer in 1908. Further study by him led to the resolution of corresponding compounds of phosphorus in 1911, and it was thought that some modified tetrahedral configuration might be adopted to discuss these amine and phosphine oxides. In any case it seemed clear that the saturation of the five valencies of a nitrogen or phosphorus atom by only four different radicles is sufficient to result in molecular dissymmetry.

Since the time when Jones adopted the pyramidal configuration for discussing the isomerism of the substituted ammonium salts, much more attention has been given to the study of different types of valency. This has come about partly in consequence of increased interest in Werner's fundamental ideas regarding the distinction between ionizable and non-ionizable valencies, and still more as the result of the elaboration of electronic theories of valency by Sidgwick and others. The idea that the optical activity of substituted ammonium salts is to be ascribed to an asymmetric ammonium *ion* as distinct from a complete molecule was gradually gaining ground. It would therefore be a simple matter to assign a tetrahedral configuration to a substituted ammonium ion of the type $[\mathrm{N}abcd]^+$. Such a view would be wholly in agreement with the modern theory that strong electrolytes are completely ionized. Direct experimental evidence in favour of a tetrahedral arrangement for ammonium ions was forthcoming in 1925 when Mills and Warren prepared and resolved 4-phenyl-4'-carbethoxybispiperidinium-1:1'-spirane bromide,

$$\left[\begin{array}{c} H \\ Ph \end{array} \!\!>\!\! C \!\!<\!\! \begin{array}{c} CH_2.CH_2 \\ CH_2.CH_2 \end{array} \!\!>\!\! N \!\!<\!\! \begin{array}{c} CH_2.CH_2 \\ CH_2.CH_2 \end{array} \!\!>\!\! C \!\!<\!\! \begin{array}{c} H \\ CO_2Et \end{array} \right]^+ Br^-.$$

The configuration of this cation might on *a priori* grounds be such that the two piperidine rings which have the nitrogen atom in common were both in the plane of the paper. In such an event there would be no molecular dissymmetry, optical isomerism would not arise, and it would be legitimate to conclude that a pyramidal arrangement for ammonium salts is correct. On the other hand, since the authors resolved the compound, the two piperidine rings must be in planes at right angles to one another, and therefore a tetrahedral configuration for the ammonium ion is the only possible one.

The stereochemistry of tervalent nitrogen received an important contribution on the theoretical side in 1890, when Hantzsch and Werner accounted for the existence of isomeric oximes, of which benzaldehyde was known to give rise to two and benzil to three of such isomerides. This theory was an extension of *cis-trans* isomerism in doubly linked carbon-to-carbon compounds, such as maleic and fumaric acids, to compounds in which a carbon atom is linked to a nitrogen atom by a double bond. One fundamental idea in the Hantzsch-Werner theory is that a tervalent nitrogen atom functions similarly to a CH group; hydrogen cyanide may be compared with acetylene or pyridine with benzene. The nitrogen atom is considered to be situated at one apex of a tetrahedron having its three valencies directed towards each of the other three apices. Adopting projection formulae the benzaldoximes would be formulated as

$$\begin{array}{ccc}
\underset{\text{Benz-synaldoxime}}{\overset{\displaystyle C_6H_5.C.H}{\underset{\displaystyle N.OH}{\|}}} & \text{and} & \underset{\text{Benz-antialdoxime}}{\overset{\displaystyle C_6H_5.C.H}{\underset{\displaystyle HO.N}{\|}}}
\end{array}$$

and the dioximes of benzil as

$$\begin{array}{ccc}
\underset{syn}{C_6H_5.C\!\!-\!\!-\!\!-\!\!-\!\!C.C_6H_5 \atop \overset{\|}{N.OH}\ \overset{\|}{HO.N}} &
\underset{anti}{C_6H_5.C\!\!-\!\!-\!\!-\!\!C.C_6H_5 \atop \overset{\|}{HO.N}\ \overset{\|}{N.OH}} &
\underset{amphi}{C_6H_5.C\!\!-\!\!-\!\!-\!\!-\!\!C.C_6H_5 \atop \overset{\|}{N.OH}\ \overset{\|}{N.OH}}
\end{array}$$

This theory was received on the whole fairly generally, as it accounted for a very considerable amount of the experimental facts. But some chemists attempted to discuss the problems of isomerism on structural as opposed to stereochemical considerations, as was originally attempted by Beckmann in 1889, the discoverer of the second oxime of benzaldehyde.

In 1910 the correctness of the Hantzsch-Werner theory was verified in an elegant manner by Mills and Bain, who showed that the three valencies of the nitrogen atom do not lie in one plane. They prepared the oxime of *cyclo*hexanone-4-carboxylic acid,

$$\underset{HO_2C}{\overset{H}{\diagdown}}C\underset{CH_2.CH_2}{\overset{CH_2.CH_2}{\diagup\diagdown}}C\!=\!N.OH, \text{ and found that this acid forms both}$$

dextro- and laevo-rotatory salts after resolution with morphine or quinine. It is clear that this optical isomerism must be due to molecular dissymmetry, and that the hydroxyl group must be in a plane at right angles to that of the hexamethylene ring, and the two enantiomorphous forms must be such that the hydroxyl group is either in a *cis* or *trans* position with respect to the carboxyl group.

In 1894, Hantzsch extended the idea of *cis-trans* isomerism to doubly linked nitrogen atoms in order to discuss the constitution of the aromatic diazo compounds. His ideas were not accepted until after a prolonged controversy with Bamberger, who attempted to explain the existence of the isomeric potassium diazotates on structural grounds in much the same way as Beckmann had unsuccessfully attempted to do in the case of the isomeric oximes. According to Hantzsch the unstable *syn* potassium diazotate has the *cis* formula $\begin{matrix} C_6H_5.N \\ \| \\ KO.N \end{matrix}$, and the stable potassium *anti*-diazotate has the *trans* formula $\begin{matrix} C_6H_5.N \\ \| \\ N.OK \end{matrix}$. It has been remarked that the experimental evidence which Hantzsch at first produced in support of his views was extremely scanty, but further experimental evidence in favour of their essential correctness was soon forthcoming. Without entering into detail, it should be noted that part of the success which Hantzsch achieved over his rival, Bamberger, was due to the fact that whereas the latter adhered to 'purely chemical' methods, Hantzsch was able to bring forward evidence of a physico-chemical character, such as the electrical conductivity of the solutions, in support of his views.

The question of the distribution in space of the three valencies of a tervalent nitrogen atom has been frequently studied by attempts to resolve compounds of this kind. One early example of such attempts was made in 1904 by Jones and Millington, who failed to resolve phenylbenzylhydrazine, $C_6H_5(C_7H_7)N.NH_2$, with the aid of *dextro*-camphorsulphonic acid. The lengthy and impressive list of failures which has been accumulated in the course of years might be regarded as evidence that the three valencies of tervalent nitrogen *can* assume a coplanar configuration. Other explanations are, however, more probable. Thus it has been pointed out by Shriner, Adams, and Marvel (Gilman's *Organic Chemistry*, I, 329) that the attempts of some of the earlier investigators, who worked with secondary amines, such as ethyl-benzylamine, mono-methylaniline, and tetrahydropyridine, were foredoomed to failure because such compounds would form salts with acids having symmetrical cations of the ammonium type which are obviously non-resolvable. It seems by no means unlikely that attempts to resolve the potentially active compounds may be vitiated by racemization. This was the view expressed by Meisenheimer in 1924, who pointed out that racemization involving the application of a very small amount of energy to the molecule might very readily arise with tervalent nitrogen.

In bringing this brief account of the stereochemistry of nitrogen to a close, it may be remarked that it is possible to bring the amine oxides into satisfactory line with a tetrahedral configuration for the asymmetric nitrogen atom. It has been pointed out by Sidgwick (*The Electronic Theory of Valency*, 1927, pp. 71, 220) that in these compounds the oxygen atom is united to the nitrogen atom by a co-ordinate link, the nitrogen atom donating two electrons to the oxygen atom, the organic groups being attached by covalencies, thus $\begin{matrix} CH_3 \\ C_2H_5 \\ C_6H_5 \end{matrix}\!\!>\!\!N\rightarrow O$. The linkage between the nitrogen and oxygen atoms is not a double bond, as Meisenheimer thought in 1908.

Optical activity due to an asymmetric sulphur atom was realized as early as the year 1900 by Pope and Peachey, and independently by Smiles. The compounds are sometimes known as thetines, or better, for reasons which will appear presently, as thionium salts. The compound resolved by Pope and Peachey was the camphorsulphonate of methyl ethyl thetine and was considered to contain quadrivalent sulphur and have the formula $\begin{matrix} CH_3 \\ C_2H_5 \end{matrix}\!\!>\!\!S\!\!<\!\!\begin{matrix} CH_2.CO_2H \\ X \end{matrix}$. All these compounds contain three organic groups attached to the sulphur atom, the fourth group X being a strongly electronegative anion, such as camphorsulphonate or bromocamphorsulphonate. As the optical activity persists irrespective of the ionization, it seems clear that the original mode of formulating these compounds requires modification. Twenty-five years later examples of derivatives of a compound having only three organic groups directly united to a sulphur atom, and showing optical activity, were described by Phillips. These were esters of toluene sulphinic acid, and according to that author should be formulated with a semi-polar double bond or co-ordinate link between the sulphur and one of the oxygen atoms to account for the asymmetry $O\leftarrow S\!\!<\!\!\begin{matrix} C_7H_7 \\ O.C_2H_5 \end{matrix}$. Between 1925 and 1927 Kenyon and Phillips with their collaborators followed this up by resolving two other types of compounds having only three groups attached to the sulphur atom, namely, sulphoxides, and the sulphilimines discovered in 1922 by Mann and Pope. These latter compounds are formulated thus:

$$C_7H_7.SO_2.N\leftarrow S\!\!<\!\!\begin{matrix} C_6H_4.CO_2H \\ CH_3 \end{matrix} \quad,$$

the sulphur atom being linked to the nitrogen atom by a semi-polar double bond as in the case of the sulphinic esters and the sulphones. The spatial arrangement would thus appear to consist of a tetrahedron with the sulphur atom at one corner and the organic groups at the other three corners.

Numerous other elements have in one way or another been concerned with the formation of asymmetric molecules, of which only a brief reference can be made to one or two interesting examples. About the time when Pope had prepared his optically active sulphur compounds, he was able to prepare similar optically active compounds of selenium, and also to obtain compounds with asymmetric atoms of tin. In all such compounds the structure was fundamentally tetrahedral. In 1907 the first example of an optically active silicon compound was prepared by F. S. Kipping. This was a compound of an ether type having two asymmetric silicon atoms in the molecule, namely, the compound sulphobenzylethylpropylsilicyl oxide,

$$\begin{array}{cc} C_2H_5 & C_2H_5 \\ | & | \\ SO_3H.C_6H_4.CH_2.Si-O-Si.CH_2.C_6H_4.SO_3H. \\ | & | \\ C_3H_7 & C_3H_7 \end{array}$$

At a later date compounds with only one asymmetric silicon atom in the molecule were prepared.

It has long been known that boric acid forms complexes with organic compounds, many of which are very much stronger acids than boric acid itself. This has been known for upwards of half a century in connexion with the titration of the acid. In 1924 Boeseken and Meulenhoff showed that optical activity could be realized with a borosalicylic acid of the formula

in which the asymmetry is clearly due to the quadricovalent boron atom, the two salicylic acid residues being in planes at right angles to one another. The compound might be described as a borospirane, and the configuration is clearly tetrahedral.

In the course of his extensive studies on isomerism in inorganic compounds, Werner found undoubted examples of *cis-trans* isomerism among compounds such as the ammines of cobalt, chromium, and platinum. He soon found that the phenomena could be discussed in a satisfactory way by considering the atom of the heavy metal to be situated at the centre of a regular octahedron, having the six co-ordinated units at the six corners. By the beginning of the present century a vast amount of material had been accumulated chiefly as the result of the activities of Werner and his pupils, and in 1912 he realized optical activity as a result of asymmetric chromium ions. Two types of such compounds were obtained. The first example contained an optically active cation, such as an ethylene-diamine compound of the type $[Cr(en)_3]^{+++} X_3^{---}$. The second was a derivative of potassium chromioxalate, $K_3^{+++} [Cr(C_2O_4)_3]^{---}$, in which the optical activity arises in consequence of the asymmetric chromioxalate anion. In these compounds the co-ordination number of six is satisfied in the case of the first-named compound by three molecules of ethylene diamine around the chromium atom; in the case of the chromioxalate the same result is brought about by the three oxalate residues around the atom of chromium. It is not easy to visualize the spatial arrangements without the use of models, but the relationship of object to mirror image in such compounds can be understood by regarding the central chromium atom having the three organic groupings attached to it in a manner similar to a triple-bladed propeller, thus

$$C_2O_4 \diagdown \diagup C_2O_4$$
$$Cr$$
$$C_2O_4$$

In 1927 Wahl and Andersin prepared and resolved ammonium alumino-oxalate, $(NH_4)_3[Al(C_2O_4)_3]$, in which the asymmetry is of a precisely similar type.

The term *steric hindrance*, or as it is sometimes called, *proximity effect*, denotes impedance of reactivity shown by certain kinds of groupings due to the presence of other groups in the vicinity. The earliest experiments bearing on this subject were carried out by Menschutkin in 1879, who investigated the effect of constitution upon the ease with which different types of alcohols are esterified. His method consisted in heating equivalent quantities of primary, secondary, and tertiary alcohols with acetic acid for one hour at a fixed temperature. In his experiments he found that the yield of ester from primary alcohols was about 46 per cent, from secondary alcohols it varied from about 17 to 27 per cent, and from tertiary alcohols the yield was from 1 to 2 per cent. The molecular weight of the alcohols had remarkably little effect upon the yield of ester, but it is noteworthy that methyl alcohol gave an abnormally high yield, viz. 56 per cent. Whatever theoretical views may be entertained regarding mechanism of these reactions, it is clear that the hydroxyl group is most reactive in primary alcohols, less so in secondary alcohols, and least reactive in tertiary alcohols, and the impedance can be put down to the spatial effect of univalent groups in the vicinity of the hydroxyl group.

In 1894, Victor Meyer began an elaborate series of experiments on the effect of introducing substituents into one or both ortho positions to a carboxyl group in an aromatic nucleus as regards the capacity of the acid for being esterified. The rate of esterification of benzoic acid when *one* ortho position is substituted by certains groups is appreciably reduced: when *both* ortho positions are substituted in this way the rate of esterification is enormously reduced. The effect of the double ortho substitution is very great, and is by no means restricted to esterification reactions. Thus Kehrmann in 1888 observed that the capacity of ketones for yielding oximes was greatly influenced by the presence of certain substituents in the ortho position to the carbonyl groups. He found that tri-substituted quinones yielded oximes with great difficulty, while tetra-substituted quinones, such as chloranil, gave no yield of oxime whatever. And in 1900 it was shown by Fischer and Windaus that anilines having both ortho positions blocked, while they formed tertiary bases easily, would not yield quaternary ammonium salts. Numerous other examples of a similar kind have been accumulated.

In attempting to explain phenomena of this kind on the basis of steric hindrance as originally understood there are very considerable difficulties. Even in Victor Meyer's time anomalies were not unknown. Coming to more modern times, Auwers and

Harres in 1929 found that 3-chloro-4-methyl benzoic acid resists esterification. In considering examples of this kind it must not be forgotten that while the usual direct method of esterification may fail to yield positive results, other methods may result in success. Thus immediate esterification is frequently effected by treating the silver salt of the acid with an alkyl iodide. The whole subject abounds with difficulties, of which particular importance must be attached to the fact that the *nature* of a substituent frequently exerts an influence altogether irrespective of its *size*. Thus the effects exerted by nitro groups are frequently similar to those exerted by methyl groups when these replace hydrogen atoms in the ortho positions to carboxyl groups. Examples are even known in which the reactivity of a group may be intensified by the presence of certain substituents in its vicinity. Thus it has long been known that the presence of electronegative atoms or groups in the molecule of an organic compound can give rise to an increase of chemical reactivity. The increase in strength of acetic acid caused by the successive substitution of the hydrogen atoms in the molecule by halogen atoms was studied by Ostwald in 1889. Similarly, the acidic character of phenols is intensified by the presence of nitro groups in the vicinity of the hydroxyl group. But when the subject is studied closely, it is found to be highly complicated. Thus in dealing with the problem of the esterification of aromatic acids having the ortho positions blocked, and of the hydrolysis of esters of this type, some exceptional examples have been encountered. Although in general a parallelism between hydrolysis and esterification was found to exist, Kellas in 1897 observed that whereas the nitrobenzoic acids were more slowly esterified than benzoic acid itself, the esters of the nitrobenzoic acids were very much more rapidly hydrolysed. This subject has been studied more elaborately by Kindler between 1926 and 1928. Of the numerous modern researches on the general effects of substituents on the reactivity of particular groups, reference may be made to an important investigation on the formation and stability of substituted aromatic aldehydic and ketonic cyanhydrins by Lapworth and Manske in 1928, who obtained some results in direct opposition to what might be expected from the standpoint of the classical spatial considerations.

In 1885, Baeyer put forward a theory to account for differences in stability of ring compounds according to the number of atoms in the rings. This theory is really an extension of the tetrahedral conception of the carbon atom as expressed by van't Hoff and Le Bel. The angle between the valencies of a carbon atom situated at the centre of a regular tetrahedron can be calculated to be 109° 28'. The instability of a ring is measured by the amount of deflexion of the valency bonds from the unstrained angle of 109° 28'. Thus in an ethylenic ring the strain is measured by an angle of 54° 44'. In trimethylene, tetramethylene, and pentamethylene compounds the amounts of strain as expressed in this way are 24° 44', 9° 44', and 0° 44' respectively. In a hexamethylene ring the strain is in the opposite sense, and equal to −5° 16'. In this way Baeyer was able to account for the relatively stable character of ring compounds of the pentamethylene and hexamethylene class as contrasted with compounds having more than six or fewer than five atoms in the rings. The theory also accounts in a simple way for the readiness with which lactones are formed from γ and δ hydroxy acids, and for the ease with which dibasic aliphatic acids of the succinic and glutaric type yield anhydrides on heating.

In 1890, Sachse extended Baeyer's original conception to considering the atoms in a ring as not being all in one plane. Among the interesting consequences of this, he put forward the idea of strainless hexamethylene rings. Although the subject has developed on somewhat different lines to Sachse's original ideas, it is now recognized that large rings of considerable stability can be obtained in which the element of strain is greatly reduced, or altogether absent.

Very important developments regarding strain in cyclic compounds have followed as the result of an idea known as the valency deflexion hypothesis of J. F. Thorpe and C. K. Ingold. These investigators have since 1915 carried out numerous experiments on ring formation, and one of their fundamental postulates is that the normal angle of 109° 28' between two valencies of a carbon atom can be altered according to the nature of the groups attached to the other two carbon atoms. Thus Ingold and Thorpe have considered that ring formation with a disubstituted glutaric acid, $\begin{matrix} R \searrow \\ R \nearrow \end{matrix} C \begin{matrix} \nearrow CH_2.CO_2H \\ \searrow CH_2.CO_2H \end{matrix}$, which can take place more readily than with glutaric acid itself,

$$\begin{matrix} H \searrow \\ H \nearrow \end{matrix} C \begin{matrix} \nearrow CH_2.CO_2H \\ \searrow CH_2.CO_2H \end{matrix} ,$$

is due to diminution of the size of the angle between the two —$CH_2.CO_2H$ groups by the presence of the groups R.

In another direction, exceptional interest is attached to the work of Ruzicka since 1926 on the preparation of large carbon rings. It had long been known that cyclic ketones, such as *cyclo*pentanone, *cyclo*hexanone, and *cyclo*heptanone, could be prepared by the dry distillation of the calcium salts of the appropriate dibasic acids. Experiments to obtain larger rings presented great difficulties, but Ruzicka and his collaborators found that the yield of these cyclic ketones was considerably influenced by the nature of the salt of the dibasic acid which is used for the preparation. In particular, he found that high yields were obtained when thorium salts were heated, and further, rings containing upwards of thirty carbon atoms could be prepared by the action of heat upon the thorium salts of the higher dibasic acids. The marked stability of these rings indicates that they must be strainless and almost certainly multiplanar. Special interest is attached to the study of these large rings, as compounds containing them have been shown to occur in nature.

It has long been known that the magnitude and sometimes even the sign of the rotation of a dissolved optically active substance may be altered by change in the physical conditions. So long ago as the year 1846 it was observed by Dubrunfaut that the rotatory power of a freshly prepared solution of glucose gradually falls and eventually attains a constant value. Other examples of a similar kind were soon forthcoming, and this subject has received much attention in modern times, notably at the hands of Lowry and his pupils since 1899, who devised the term *mutarotation* to characterize these phenomena. Bearing in mind that any alteration in the magnitude or sign of the rotatory power of a dissolved substance is a clear indication of the occurrence of chemical change, it will be helpful to refer back to Pasteur's experiments about the year 1853 in which he found that *dextro*-tartaric acid could be converted into a mixture of racemic and *meso*-tartaric acids by the action of heat. This seems to be the first recorded example of *racemization,* viz. the conversion of an optically active compound into an inactive modification. Since Pasteur's time racemization has been studied from several points of view, and here a passing reference may be made to the remarkable transformations which Fischer effected about the year 1890 in the sugar group. Fischer found that gluconic and mannonic acids, which are identical in configuration except as regards the carbon atom adjacent to the carboxyl group, were interconvertible when either acid was heated with pyridine or quinoline. In 1895 a discovery of exceptional interest was made by Walden, which has since become known as the *Walden inversion.* This consists in the reversal of the sign of rotation of an optically active compound when a substituent attached to an asymmetric atom is exchanged for another one. Perhaps the best known example of a Walden inversion, and one of the earliest to be studied, arises in connexion with the transformation of chlorosuccinic acid into malic acid and vice versa, and may be represented as follows:

$$\begin{array}{ccccc} d\text{-chlorosuccinic acid} & \xrightarrow{\text{Ag}_2\text{O}} & d\text{-malic acid} \\ \text{KOH} \swarrow\nwarrow \text{PCl}_5 & & \text{PCl}_5 \swarrow\nwarrow \text{KOH}. \\ l\text{-malic acid} & \xleftarrow{\text{Ag}_2\text{O}} & l\text{-chlorosuccinic acid} \end{array}$$

It will be evident that silver oxide and potassium hydroxide behave differently in replacing the chlorine atom by a hydroxyl group. Silver oxide effects substitution without causing any change in the sign of the rotation, whereas with potassium hydroxide reversal takes place. The reconversion of malic acid into chlorosuccinic acid by phosphorus pentachloride takes place with reversal of the rotation.

Between 1907 and 1911, Fischer made an elaborate study of optical inversion in the course of his studies on amino acids. Thus he found that when the bromine atom in an α-bromo-carboxylic acid is replaced by an amino group by the action of ammonia, an amino acid of similar rotatory power is obtained, and that this is the case whether the free acid or one of its esters is employed. But if the reaction is reversed, the relations are less simple. Thus when alanine (α-aminopropionic acid) is treated with nitrosyl bromide, an α-bromopropionic acid with the optical rotation reversed is produced, whereas if the ethyl ester of alanine is treated in this way and then hydrolysed, the formation of the α-bromopropionic acid takes place without reversal of the rotation. Although this behaviour has been observed in other cases, an important exception was found with valine, α-amino*iso*valeric acid. When *l*-valine is treated with nitrosyl bromide it produces *d*-α-bromo*iso*valeric acid, but this compound when treated with ammonia regenerates *l*-valine

A large amount of material has been collected as the result of the experimental activities of numerous investigators on this subject extending over many years. As far as the action of reagents which effect identical substitutions are concerned, an important difference between the action of phosphorus pentachloride and thionyl chloride has been observed by McKenzie and his collaborators in a series of researches from 1908 onwards. They have found that whereas phosphorus pentachloride replaces a hydroxyl group by a chlorine atom with reversal of optical rotation, the use of thionyl chloride in general effects the replacement without such reversal. Thus *l*-ethyl atrolactinate (the ethyl ester of methylphenylhydroxyacetic acid), $\begin{smallmatrix} C_6H_5 \\ \\ CH_3 \end{smallmatrix}\!\!>\!\!C\!\!<\!\!\begin{smallmatrix} OH \\ \\ CO_2Et \end{smallmatrix}$ is converted by phosphorus pentachloride into *d*-phenylchloroacetate, while thionyl chloride effects conversion into the *l*-phenylchloroacetate. This difference in behaviour between the two chlorides will bear comparison with that observed between silver oxide and potassium hydroxide.

In attempting to obtain some insight into the bewildering mass of experimental facts which has been collected, it may be helpful to consider some studies which have been directed more particularly towards the phenomena of racemization. Much attention has been given to the fact that racemization and also mutarotation are to be observed particularly in compounds which have a mobile hydrogen atom in the molecule. Thus in 1893, Fischer discovered two isomeric methyl derivatives of glucose which differed in rotatory power, but showed no mutarotation. Two modifications of glucose were discovered by Tanret in 1896. Both varieties of glucose displayed mutarotation, but eventually the same final value is obtained, which is that of the equilibrium mixture of both sugars. Further, it was found possible to prepare both α-glucose and β-glucose from the corresponding methyl glucosides. Since 1899 this subject has received a great deal of attention, and while opinion has varied regarding the *mechanism* of the isomeric changes concerned in the transformations taking place between isomeric sugars, there is no doubt regarding the interference with such chemical changes by the replacement of the mobile hydrogen atom by a methyl group. Similar considerations have been established by the researches of McKenzie and his pupils since 1908 regarding the types of optically active acids which are or are not readily racemized. Thus it was found that mandelic acid and phenylchloroacetic acids and their esters, in which a hydrogen atom is directly attached to the asymmetric carbon atom, are easily racemized. On the other hand, the corresponding acids and their esters in which the hydrogen atom has been replaced by a methyl group, viz. atrolactinic (methyl mandelic) acid and methyl-phenylchloroacetic acid, are not racemized.

More recently, F. B. Kipping in 1935 published an interesting study of certain sulphonylthioethanes. In these compounds the structure which gives rise to optical activity consists of an asymmetric carbon atom with the groupings $\begin{matrix} R.SO_2 \\ Me \end{matrix} >C< \begin{matrix} H \\ S.C_6H_5 \end{matrix}$. He succeeded in resolving the externally compensated α-p-carboxyphenylsulphonyl-α-phenylthioethane,

$$CO_2H.C_6H_4.SO_2.CH(Me).S.C_6H_5,$$

and found that the methyl esters of this and of other closely similar compounds were easily racemized under the influence of alkalis. Kipping concluded that the racemization was associated with the mobile hydrogen atom attached to the asymmetric carbon atom; and he found that if this hydrogen atom was replaced by a phenyl group, the resulting compound was not easily racemized.

It must, however, be added that explanations based upon the presence of a mobile hydrogen atom in the molecule are not always applicable. Two acids are known which, differing solely by one containing an iodine atom and the other an atom of bromine in the molecule have been found to show marked differences in regard to racemization. Chloroiodomethane-sulphonic acid, studied by Pope and Read in 1914, was shown to resist the action of racemizing agents, whereas chlorobromomethane-sulphonic acid was examined by Read and McMath in 1925 and found to be racemized with considerable rapidity.

In 1914 an elegant experimental proof of a fundamental problem in stereochemistry was given by Fischer and Brauns. They succeeded in interchanging two groups attached to the same asymmetric carbon atom in a compound so as to cause optical inversion by a series of changes in which all the intermediate compounds were isolated. The value of the work is due to the fact that the result could not be vitiated by a Walden inversion because the groups concerned remained attached to the asymmetric carbon throughout the transformations. Starting with d-isopropylmalonamic acid, $\text{C}_3\text{H}_7\text{—C(H)(CO.NH}_2)(\text{CO}_2\text{H})$, this was converted by diazomethane into the methyl ester, $\text{C}_3\text{H}_7\text{—C(H)(CO.NH}_2)(\text{CO}_2.\text{CH}_3)$, and thence by the action of nitrous acid into d-isopropylmalonic methyl ester, $\text{C}_3\text{H}_7\text{—C(H)(CO}_2\text{H})(\text{CO}_2.\text{CH}_3)$. Then by the action of hydrazine, the l-isopropylmalonyl hydrazide acid, $\text{C}_3\text{H}_7\text{—C(H)(CO}_2\text{H})(\text{CO}.\text{N}_2\text{H}_3)$, was obtained, and this compound converted by nitrous acid into l-isopropylmalonylazide acid, $\text{C}_3\text{H}_7\text{—C(H)(CO}_2\text{H})(\text{CO}.\text{N}_3)$. Finally, by the action of ammonia followed by hydrochloric acid, hydrazoic acid was eliminated with formation of l-isopropylmalonamic acid, $\text{C}_3\text{H}_7\text{—C(H)(CO}_2\text{H})(\text{CO}.\text{NH}_2)$. The values obtained for the rotatory power of the original dextro-acid and of the final laevo-acid were sufficiently close in magnitude and opposite in sign to show clearly that the interchange of the two substituents in this asymmetric system had been effected. It will have been noted that at no stage of these transformations has any group directly attached to the asymmetric carbon atom been completely removed.

In a most interesting series of studies on the stereochemical aspects of displacement reactions, Kenyon and Phillips with their collaborators have, since 1923, found that certain secondary alcohols when acetylated directly by using acetic anhydride yield products of the same sign of rotation as the original alcohol, whereas when the acetyl compound is prepared indirectly *via* the *p*-toluenesulphonate followed by the action of potassium acetate, it has an equal rotatory value but of opposite sign. The transformations may be summarized as follows:

$$
\begin{array}{ccc}
\underset{dextro}{R\!\!>\!\!C\!\!<\!\!\overset{OH}{H}} & \xrightarrow{\;C_7H_7SO_2Cl\;} & \underset{dextro}{R\!\!>\!\!C\!\!<\!\!\overset{O.SO_2C_7H_7}{H}} \\
(CH_3CO)_2O \downarrow & & \downarrow CH_3CO_2K \\
\underset{dextro}{R\!\!>\!\!C\!\!<\!\!\overset{O.CO.CH_3}{H}} & & \underset{laevo}{R\!\!>\!\!C\!\!<\!\!\overset{H}{O.CO.CH_3}}
\end{array}
$$

it will be clear that the only possibility of inversion must have arisen in connexion with the rupture of one of the linkages of the asymmetric carbon atom, and that this can only have occurred in the transformation of the *p*-toluenesulphonate into the acetate.

The kinetics of reactions concerned with racemization and optical inversion have attracted the attention of numerous chemists over many years; indeed, some of the researches date back almost to the discovery of the phenomena. In recent years, especially about 1937, the Walden inversion has been the subject of an elaborate study by Ingold and Hughes with their collaborators. Among the numerous reactions which they have studied, mention may be made of the hydrolysis of halogen-substituted compounds, such as bromopropionic acid. This acid and its methyl ester were found to undergo hydrolysis by a reaction of the second order, which takes place with inversion and without appreciable racemization. The hydrolysis of sodium bromopropionate was found to proceed by both types of mechanism, namely, first and second order reactions, the former giving almost complete retention of the original configuration and the latter taking place with inversion. The general conclusion which has been established by the work of Ingold and Hughes, especially with the replacement of halogens in alkyl halides, is that hydrolysis or alcoholysis which proceeds according to reactions of the second order does so with inversion of configuration but without any racemization; whereas when hydrolysis or alcoholysis of these types of compounds takes place by reactions of the first order, considerable racemization takes place accompanied with inversion. These results are of much interest, not only from the standpoint of stereochemistry, but from the modern views on the general subject of chemical kinetics. It will be seen (Chapter VIII) that the distinction between reactions of the first and second order, or as they are frequently termed, unimolecular and bimolecular reactions, is no longer regarded as fundamental but the difference is one of degree.

It is interesting to recall that the subject first studied by Pasteur nearly a century ago, namely, crystallography, led to the foundation of the whole science of stereochemistry. The history of the relations between crystallography and chemistry since the early period, which began with Berzelius and Mitscherlich, to modern times is an interesting example of a subject which attracted much attention at first, then after a period of neglect assumed increased importance in more recent years. The chemical aspect of the subject which was of first importance in the Berzelius-Mitscherlich period was undoubtedly that of atomic weight determinations. Thus it was a comparatively simple problem to settle the atomic weight of an element such as selenium on account of the isomorphism of salts such as potassium sulphate and potassium selenate. At that time it was assumed that isomorphism in the strictly chemical sense of the capacity to form mixed crystals implied identity of valency. A few apparent exceptions were indeed noted, and this may possibly have had something to do with the way in which many chemists were inclined to neglect crystallographic studies. The second period may be termed the valency period. So long as problems of structural theory were not under consideration studies in isomorphism continued to be of interest to chemists, but once it had been recognized that two compounds so dissimilar as calcium carbonate and sodium nitrate were truly isomorphous it was considered that such a result was altogether irreconcilable with structural theory. A notable attempt was made by Barlow and Pope in 1906 to correlate crystallographic properties with valency, and although their theoretical ideas were subsequently shown to be unsound, particularly with regard to inorganic compounds, the interest of chemists in crystallography was largely revived as a result of their studies. Barker in 1912 after pointing out some of the fallacies of the Barlow-Pope theory showed that much order and regularity could be introduced into the subject by way of Werner's co-ordination theory. Thus while the isomorphism of sodium nitrate and calcium carbonate was meaningless on the older theories, it became intelligible by writing the formulae of the compounds as $Na[NO_3]$ and $Ca[CO_3]$, the oxygen atoms being arranged in a similar manner around the nitrogen or carbon atoms respectively.

A new chapter was opened in 1912 by Laue and his collaborators, Friedrich and Knipping, who showed that crystals can diffract X-rays. This discovery was immediately followed up by the Braggs, who developed the subject of X-ray analysis by using crystals as reflexion gratings, and later in 1916 by Debye and Scherrer, who devised a diffraction method using powdered crystals. The results of these methods were of immediate importance to chemists. One of the first fruits of the work of the Braggs was the elucidation of the structure of the diamond. They found that each carbon atom throughout the whole structure is immediately surrounded by four others in a regular tetrahedral manner—surely a most vivid verification of the theory of van't Hoff and Le Bel.

In some directions the modern methods of X-ray analysis and of electron diffraction have not only confirmed but surpassed results obtained by purely stereochemical considerations. The idea of relative distances between atoms within molecules was tacitly assumed by van't Hoff and by Wislicenus in developing the conception of *cis-trans* isomerism, and the same may be said with regard to Kekulé's conception of position isomerism in aromatic compounds. This subject was certainly regarded in this way by Körner between 1867 and 1874 in connexion with the problem of orientation, the ortho, meta, and para positions being considered from this fundamental point of view. By the methods of X-ray analysis, however, it is now possible to discuss atomic distances in numerical terms. The nature of the allotropy of the two crystalline forms of carbon has been elegantly determined by these methods. While the Braggs in 1913 found that the atomic structure of diamond is fundamentally tetrahedral, it was shown by Hassel and Mark and independently by Bernal in 1924 that the structure of graphite is hexagonal. As the result of the work of the numerous investigators in this field, it is now established that while diamond is the prototype of aliphatic compounds, the prototype of aromatic compounds is undoubtedly graphite. The carbon-to-carbon distances, expressed in Ångstrom units, are now recognized to be for diamonds and for singly linked aliphatic compounds 1·54, for ethylenic linkages 1·33, for acetylenic linkages 1·20, and for graphite and many aromatic compounds 1·42. This value for aromatic compounds has an interesting bearing on the formula for benzene. Ever since the time of Kekulé the formula of this fundamental aromatic substance has been under discussion, particularly as regards the disposition of the fourth carbon valency. The idea of some kind of intra-annular tautomerism which was favoured for many years is no longer popular, and an idea of *resonance* in which the molecule cannot be expressed by a single formula, but is an intermediate of several possible forms, is becoming more widely accepted, and one important line of evidence in favour of such a view is that a value of 1·42 Ångstrom units does not correspond to any one recognized type of atomic linkage.

REFERENCES

Louis Pasteur. *Researches on the Molecular Asymmetry of Natural Organic Products* (1860). Alembic Club Reprints, No. 14. See also *Œuvres de Pasteur*, Paris, 1922.

J. H. van't Hoff. *The Arrangement of Atoms in Space*, with a preface by J. Wislicenus, and an appendix on Stereochemistry among Inorganic Substances, by A. Werner. London, 1898.

Percy Frankland. Presidential Address to the Chemical Society. *J. Chem. Soc.* 1912, p. 654.

M. O. Forster. Emil Fischer Memorial Lecture. *J. Chem. Soc.* 1920, p. 1157.

G. Wittig. *Stereochemie*. Leipzig, 1930.

K. Freudenberg. *Stereochemie*. Wien, 1933.

W. H. Mills. Presidential Address to the Chemistry Section of the British Association. *Chemistry and Industry*, 1932, p. 750.

P. D. Ritchie. *Asymmetric Synthesis and Asymmetric Induction*. St Andrews University Publication, No. xxxvi, 1933.

W. Hückel. *Theoretische Grundlagen der Organischen Chemie*. Two volumes. Leipzig, 1931.

T. M. Lowry. *Optical Rotatory Power*. London, 1935.

W. H. Bragg and W. L. Bragg. *X-rays and Crystal Structure*. London, 1915. Fifth edition, 1925.

Annual Reports of the Chemical Society since 1904.

Chapter IV

RADIOACTIVITY

The discovery of radioactive phenomena dates from the year 1896, when Henri Becquerel was engaged upon investigations on the fluorescent properties of uranium salts under the influence of X-rays which had been discovered very shortly before. Becquerel soon found that the fluorescent and phosphorescent properties of uranium salts were wholly unconnected with their previous history as regards exposure to any form of radiation, and further that other characteristics of these compounds, such as their property of affecting photographic plates and of causing the air in the vicinity to acquire the property of conducting electricity, were fundamentally connected with the element uranium.

Before following the rapid developments of the subject immediately after Becquerel's discovery, it is desirable to review very briefly some of the phenomena which occur when electric discharges are passed through gases at extremely low pressures, since the nature of the rays emitted by radioactive substances are closely similar to the types of radiation encountered in the passage of electricity through gases. The fundamental experiments which led to the recognition of the cathode rays were made as long ago as the year 1858 by Plücker, and in 1869 by Hittorf. Other experiments on the same lines were carried out by Varley in 1871, and by Crookes in 1879, who is usually quoted as the discoverer of the cathode rays as he was certainly responsible for their study in greater detail than the earlier workers, and possibly because his ideas regarding matter in an ultra-gaseous or *radiant* state arose from these experiments.

On this subject, some views expressed by Lord Kelvin in 1897 are of interest: 'Varley's fundamental discovery of the cathode torrent, splendidly confirmed and extended by Crookes, seems to me to necessitate the conclusion that resinous electricity, not vitreous, is *the electric fluid*, if we are to have a one-fluid theory of electricity. Mathematical reasons...prove that if resinous electricity is a continuous homogeneous liquid, it must...be endowed with a cohesional quality....It is just conceivable...that this idea may deserve careful consideration. I leave it, however, for the present, and prefer to consider an atomic theory of electricity foreseen as worthy of thought by Faraday and Clerk Maxwell, very definitely proposed by Helmholtz in his last lecture to the Royal Institution, and largely accepted by present-day theoretical workers and teachers. Indeed Faraday's law of electro-chemical equivalence seems to necessitate something atomic in electricity, and to justify the very modern name *electron*. The older, and at present even more popular, name *ion* given sixty years ago by Faraday, suggests a convenient modification of it, *electrion*, to denote an atom of resinous electricity. And now, adopting the essentials of Aepinus' theory, and dealing with it according to the doctrine of Father Boscovich, each atom of ponderable matter is an electron of vitreous electricity; which with a neutralizing electrion of resinous electricity close to it, produces a resulting force on every distant electron and electrion which varies inversely as the cube of the distance, and is in the direction determined according to the well-known requisite application of the parallelogram of forces' (*Mathematical and Physical Papers*, VI, 145).

It will have been noted that Kelvin expressed a preference for 'an atomic theory of electricity'. This is of great interest because many of the developments of electrical science in the nineteenth century had been discussed successfully in terms of electricity regarded as a fluid or as some sort of continuum. The ideas of Crookes, although now regarded as erroneous, were of value, as he considered the cathode rays to consist of material particles not of aethereal waves, as some Continental physicists, e.g. Lenard, supposed. Crookes, however, regarded the cathode-ray particles as having ordinary atomic dimensions. In 1897, J. J. Thomson and his pupils showed that the mass of these particles is about 1/1840 of that of the hydrogen atom. It was known that the cathode-rays could be deviated by a magnetic field, and Perrin showed that the particles are negatively charged. The determination of the ratio of the charge to the mass—one of the greatest triumphs of modern experimental physics—was effected by J. J. Thomson by the application of magnetic and electrostatic fields to the cathode-rays. This work provided a final and convincing experimental demonstration of the discontinuous structure of electricity. But it is interesting to note that some writers on theoretical physics, Lorentz and Larmor for example, had assumed the existence of isolated electric units, and had used Johnstone Stoney's term electron shortly before this time.

The relation between the cathode-rays and the very penetrating X-rays discovered by Röntgen was soon recognized. The latter rays are produced by the intense negative acceleration suffered by the cathode-rays in impinging upon some very dense object. This subject, as will be seen later, is relevant to the study of certain types of radiation emitted by radioactive substances.

Besides the cathode-rays and the X-rays, a few words must be added about a third type produced in discharge tubes, namely, the canal-rays, so called because their discovery by Goldstein in 1886 was facilitated by the use of perforated cathodes. These rays were shown by Perrin in 1895 to be positively charged, and in 1898 Wien determined the ratio of the charge to the mass by the method of applying magnetic and electrostatic fields. The results obtained showed clearly that, unlike the cathode-rays, the positive-ray particles are of atomic dimensions. In the hands of J. J. Thomson and afterwards of Aston these positive rays have become of very great importance in connexion with the study of isotopes, as will be seen in Chapter v.

Very considerable advances have been made in the study of the effects produced by different kinds of rays as the result of experiments carried out by C. T. R. Wilson on the condensation of moisture in a gas supersaturated with water vapour, and rendered conducting by exposure to some ionizing agency. Aitken had shown in 1880 that air supersaturated with moisture would not necessarily produce a fog unless dust particles were present to act as nuclei for the condensation of minute drops. In 1897, Wilson showed that gaseous ions act in a precisely similar manner to dust particles in functioning as nuclei for the formation of clouds. In 1911, Wilson extended the possibilities of his discovery, and devised an apparatus for obtaining visible tracks of the rays emitted from radioactive substances, by photographing the mists produced by the passage of the rays through air supersaturated with moisture. These methods have been of the greatest value in some of the most recent developments of radioactive research.

An important step in the development of the study of electrons was the proof of the identity of the magnitude of the negative charge on a gaseous ion with that carried by a hydrogen atom in electrolysis. This identity was first shown by Townsend in 1897–8 who worked on the mists obtained in electrolysis, and he obtained a value for e of 5×10^{-10} electrostatic unit. A somewhat similar figure was obtained at about the same time by J. J. Thomson who utilized C. T. R. Wilson's method of producing clouds in gases ionized by X-rays or by ultra-violet light. The weight of the cloud was estimated from the amount of water vapour required to saturate the gas and the amount in the gas before the adiabatic expansion. The number of drops per unit of volume was deduced by observing the rate at which the cloud gradually fell, and the total charge on the gaseous ions was found from the current between an electrode above the cloud and the surface of the water at the lower part of the vessel. A very similar value was obtained when the gases were ionized by Becquerel rays. Great accuracy was not to be expected from difficult experiments of this kind, but it should be noted that the results obtained by different methods varied between 3×10^{-10} and 6×10^{-10} electrostatic unit. Later, Millikan (1910–11), using oil droplets instead of clouds produced by condensation of moisture, obtained a value of 4×10^{-10} electrostatic unit for the elementary charge of negative electricity.

After the discovery of the Becquerel rays, the study of radioactive substances proceeded in two different but related directions, namely, on the physical side associated more particularly with the name of Rutherford, and on the chemical side with which the names of Madame Curie, Marckwald, Soddy, Fajans, and others are especially identified. The first work of importance carried out by Rutherford and his pupils was the demonstration of the complex nature of the radiations emitted from uranium and thorium. By experiments on the absorption of the rays by matter and on the effect of powerful magnetic fields (1899–1902), it was shown that the radiations from radioactive substances, first known by the general term Becquerel rays, consisted of three definite types, which were named by Rutherford as the α-, β- and γ-rays respectively. The α-rays have very little power of penetration, are of atomic dimensions, and deviation experiments have shown them to carry a positive charge. The β-rays have considerably greater penetrative power, are of electronic dimensions, readily deviated by a magnetic field, and are negatively charged. They are, indeed, closely similar to cathode-rays, but have in general a higher velocity. The γ-rays are characterized by very great penetrative power; thus they can traverse several centimetres of lead without serious absorption, are undeviated by a magnetic field and are of the nature of X-rays. It may be remarked that at that time, the nature of X-rays, and consequently of γ-rays, was somewhat imperfectly understood, but it was generally recognized that both of these types of rays were of the nature of undulations of extremely short wavelength: a conclusion which was established at a much later date especially by the researches of W. H. and W. L. Bragg. The atomic character of the α-rays has received further confirmation by Crookes (1903), who showed that the phosphorescence of a zinc sulphide screen exposed to these rays can be seen as a number of momentary flashing points of light.

Although the α-rays are endowed with only very feeble penetrating power, they are nevertheless possessed of considerable capacity for ionizing gases through which they pass. As will be seen later, this property has been turned to account in some experiments carried out in 1908 by Rutherford and Geiger, who proved conclusively that the α-particle is identical with the helium atom.

The photographic activity of radioactive substances is essentially a property of the β-rays. In the early stages of the study of the radioactivity of uranium compounds there was no clear distinction between the radiations which were responsible for the various effects which were produced; hence the name of Becquerel rays was used in a general way to characterize the phenomena. In 1900, however, it was shown by Crookes and by Becquerel that the photographic activity of uranium could be removed by appropriate chemical treatment. Thus Crookes showed that if excess of ammonium carbonate is added to a solution of a uranyl salt, the precipitated carbonate is dissolved in excess of the reagent, but on filtering the solution an almost invisible precipitate, consisting of impurities precipitated as hydroxides, is obtained which adsorbs the whole of the photographically active constituent of the uranyl salt. Thinking that he had isolated a new element which caused the radioactivity of uranium compounds, Crookes proposed the name of uranium-X for the new substance. Becquerel made numerous experiments on the precipitation of barium sulphate in solutions of uranium salts, and he found that the precipitate removed very considerable amounts of the constituent which caused the photographic activity. It may be added that Becquerel differed from Crookes regarding the explanation of the separation of the photographic activity from the solutions of uranium salts: he regarded the barium sulphate as having acquired some kind of induced activity. Later, Becquerel showed that precipitates of barium sulphate obtained in this way lost their photographic activity completely after the lapse of several months; but at the same time the uranium solutions, from which the barium sulphate had been precipitated, had fully recovered their original activity. This subject was studied more fully by Soddy in 1902, and he found that the processes of Crookes and Becquerel effected only the separation of the β-radiation from the uranium salts, while the α-radiation remained wholly unaffected. In particular the α-ray activity as measured by electroscopic methods was found to be identical before and after the salt was subjected to Crookes's method for removing the photographically active constituent. This subject was further investigated in 1902 by Rutherford and Grier who examined the radiations from uranium salts and from uranium-X in a magnetic field, and confirmed that the chemical separation had analysed the radiation into its two components without affecting the nature or the intensity of the rays in any way.

After the discovery of radioactivity as a fundamental property of uranium compounds, Madame Curie in 1898 undertook an elaborate search for other elements which might possess this interesting property. She found only one other element, viz. thorium, to be radioactive. When examined by electroscopic methods, it was found that the radioactivity of uranium and thorium was of the same order of magnitude, but when compared by a photographic plate, thorium is much less active than uranium. This is tantamount to saying that the α-rays are of about equal intensity for the two elements, but that the β-radiation from thorium is much weaker than that from uranium. In the course of the examination of numerous uranium minerals, Madame Curie found that certain pitchblendes exhibited a degree of radioactivity about three or four times as great as that of metallic uranium. Having found that the activity of uranium compounds, as measured by electrical methods, was proportional to the content of this element, Madame Curie concluded that pitchblende must contain some hitherto unknown element, much more radioactive than uranium, to account for its exceptional properties. Accordingly, she subjected large quantities of pitchblende to an elaborate separation of the constituents, following the separations by experiments with an electroscope; and she found that constituents having an intense degree of radioactivity were concentrated with two elements, namely, with barium and with bismuth. The unknown elements were termed radium and polonium respectively, the former being associated with barium and the latter with bismuth.

The fractionation of the barium chloride separated from pitchblende was supplemented by spectroscopic observations, and it was found that the activity of the preparations was very considerable long before any new lines were found in the barium spectrum. Eventually new lines did appear, and finally a preparation of radium chloride was obtained about a million times as active as a corresponding weight of uranium. Madame Curie determined the atomic weight of the new element by converting the chlorine in a known weight of the salt into silver chloride and obtained a value of 225 for the atomic weight of radium. It is interesting to note that a very careful series of determinations by Hönigschmid in 1912, using much larger quantities of radium chloride, gave a figure of 226 for the atomic weight of this element. No subsequent improvement has been made on this value. Radium thus takes its place in the second group below barium in the periodic classification.

In the course of the examination of large quantities of pitchblende residues obtained from Madame Curie, Debierne in 1899 obtained evidence of the existence of a third radioactive element among the hydroxides precipitated in the ammonia group, which he named actinium. In 1902, Giesel described a substance which he termed emanium, or the emanation substance from pitchblende, and considered it to belong to the rare earths, being particularly associated with lanthanum. There is now no doubt whatever regarding the identity of the substances described by Debierne and Giesel, but the name actinium has been retained for the element.

The strongly radioactive element polonium which Madame Curie found to be associated with bismuth in the course of the chemical examination of pitchblende residues was found to differ from the others in two respects, namely, by giving off only α-rays, and by the activity decaying with lapse of time. In 1902, Marckwald showed that polonium could be separated from bismuth extracted from pitchblende residues by immersing a rod of bismuth in a solution of the chloride. The whole of the activity is thus deposited on the rod. The new element was at first termed radio-tellurium by Marckwald, who showed that it was more akin to tellurium than to bismuth, but the name of polonium, originally given by Madame Curie, has remained in general use.

Of the five radioactive elements recognized at the beginning of the present century, namely, uranium, thorium, radium, polonium, and actinium, three are characterized by evolving gases having radioactive properties. These radioactive *emanations*, as they were originally termed, are produced from thorium, radium, and actinium, but not from uranium or polonium. Evidence of the gaseous nature of the emanation from thorium was obtained by Rutherford and Soddy in 1902, who further found that the gas was wholly inert towards chemical reagents, and was entirely unaffected by exposure to very high temperatures. Similar conclusions were reached as the result of a careful study of the radium emanation, and in 1903 they found that both emanations were completely condensed when passed through a tube cooled in liquid air. When the tube was allowed to warm up, the emanation again volatilized unchanged.

Just as Crookes and Becquerel found that the treatment of solutions of uranium salts by suitable chemical methods resulted in the separation of a constituent responsible for the photographic activity of this element, so Rutherford and Soddy in 1902 found that somewhat similar considerations were applicable to the radioactivity of thorium. They found that if thorium hydroxide was precipitated from a solution of the nitrate by the action of ammonia, the insoluble hydroxide had no β-ray activity, only 25 per cent of the α-ray activity, and no emanating power, whereas the filtrate contained the whole of the β-ray activity, 75 per cent of the α-ray activity, and all the emanating power. These results led Rutherford and Soddy to the conclusion that the greater part of the radioactivity of thorium is due to the presence of a substance which they termed thorium-X, following the nomenclature of Crookes, and which is separable from thorium by chemical means. It may be noted that this separation fails if the thorium hydroxide is precipitated by certain other reagents such as sodium carbonate. The cause of such failure is now intelligible because thorium-X belongs to the alkaline earths, and is indeed identical with radium in chemical properties. It should be emphasized that it is not thorium but thorium-X which produces the emanation.

Rutherford and Soddy in 1902–3 made careful quantitative measurements of the rates of decay of uranium-X and of thorium-X, and also of the recovery of the activity of the original substances after the chemical separations. They found that these processes follow an exponential law, the change of activity proceeding in geometrical progression with the time. For uranium-X the time for the activity to fall to half of its initial value was found to be 22 days, and for thorium-X the corresponding time was about 4 days. Later revised figures of 24·6 days and of 3·65 days were assigned to uranium-X and to thorium-X respectively. When the results were expressed graphically the complementary character of the two curves was immediately obvious. Having regard to the continuous production and destruction of radioactive matter, the fact that the processes are wholly independent of the external physical conditions and proceed according to an exponential law, and to the enormous liberation of energy involved, Rutherford and Soddy in 1903 put forward their theory of atomic disintegration to account for the phenomena. According to this theory a certain fraction of the atoms of a radioactive element become unstable and decompose spontaneously with the liberation of the familiar types of rays, the intensity of the radioactivity being very closely connected with the rapidity of decay. The reason why elements such as uranium and thorium exhibit radioactive properties coupled with apparent permanence is that in such elements it is only an infinitesimal fraction of the total number of atoms which are at any one time in an unstable condition. On the other hand, thorium-X must be regarded as a more active substance than uranium-X, and uranium-X as more active than polonium, since the times for the activity to be reduced to half of the initial values are 3·65, 24·6, and 136 days respectively.

It was observed by Madame Curie in 1899, and by Rutherford in the following year, that radioactive substances which emit emanations transmit temporary activity to objects in the vicinity. It was soon found that this kind of activity arises in consequence of material substances deposited in such extremely minute quantity as to be recognizable solely by their radioactive properties. The rate of decay of these active deposits follows an exponential law similar in all essential respects with that observed for the emanations. The phenomena are, however, complicated by the fact that these active deposits have been shown to be not single substances, but a series of *successive* disintegration products to which names such as radium A, radium B, etc., have been assigned. The theory of successive transformations is somewhat complicated and cannot be discussed here. It must, however, be noted that great variations in the nature and activity, as measured by the time for half-transformation, ranging from minutes to years, are to be found among the constituents of these active deposits.

The problem of the origin of radium occupied the attention of Soddy shortly after the promulgation of the disintegration theory. At first it was considered that radium was a disintegration product of uranium. The occurrence of radium in uranium minerals, and the respective atomic weights of the two elements, gave considerable support to this idea. But experiments carried out by Soddy between 1905 and 1910 showed that the rate of production of radium from uranium was very much slower than it should be if radium were a *direct* disintegration product of uranium. In the meantime an intermediate radioactive element, named ionium by Boltwood who discovered it in 1907 in uranium minerals, introduced a difficulty into the problem. It was soon apparent that the slow growth of radium in solutions of uranium salts could be traced to the presence of small quantities of ionium present with the uranium. There was certainly no definite proof of the formation of radium from *pure* uranium compounds. It is now recognized that the period for half-transformation of ionium is of the order of 10^5 years, whereas that of radium is only about 2000 years, and hence the non-appearance of radium from *pure* uranium preparations is intelligible.

The course of the atomic disintegration of thorium was also found to be considerably more complicated than was at first supposed. Thorium does not disintegrate directly into thorium-X, but via three intermediate products, known respectively as mesothorium 1, mesothorium 2, and radiothorium, and having periods of half-transformation of 5·5 years, 6·2 hours, and 2 years respectively. These substances were discovered by Hahn (1905–8). In the course of working up a large quantity of thorianite to extract the radium, some abnormalities were observed. It was found that the activity of the material present along with barium became concentrated at both ends of the fractionation. The active material in the more soluble portions was clearly a new element which was termed radiothorium. In the least soluble fractions it was found that in addition to radium a second new element, viz. mesothorium, intermediate between thorium and radiothorium, was present. Gradually it became apparent that the radiothorium which had been separated from the thorianite was not that present in the mineral at the beginning, but that which had been produced from the mesothorium *after it had been separated from the thorium* in the mineral. Eventually it was found that mesothorium could be analysed into mesothorium 1 having a long life which disintegrates raylessly into mesothorium 2 which is short lived and disintegrates with emission of β- and γ-rays in changing into radiothorium. In 1910–11 it was shown by Soddy and by Marckwald that radium and mesothorium 1 were chemically non-separable, and other examples of chemically non-separable elements were soon forthcoming. Thus Boltwood showed that ionium could not be separated from thorium, and more than one observer noted that uranium-X and thorium were non-separable.

The three radioactive disintegration series, namely, those of uranium, thorium, and actinium, have certain characteristics in common. Of these particular importance should be attached to the production of lead as the final product of radioactive change, and to the liberation of the inert gaseous emanations to which reference has already been made. The radium emanation, it should be added, belongs to the uranium series.

Since the time for the activity of the radium emanation to be reduced to one-half of its initial value is 3·85 days, whereas the corresponding times for the emanations of thorium and of actinium are measured in seconds, the properties of this gas have been studied in considerable detail. The emanation may be obtained by dissolving 30–50 mg. of a radium salt in water and removing the gases from solution by a pump. After purification the volume of emanation is a fraction of a cubic millimetre. Nevertheless, some of its physical constants have been determined with a very fair degree of exactness. Rutherford and Miss Brooks in 1901 attempted to determine its molecular weight by diffusion experiments, but the results were unsatisfactory. Other experimentalists attacked this problem, and in 1910 Debierne, using a modification of Bunsen's effusion method, obtained a value of about 220, which, as will be seen later, is remarkably close to the value obtained from a consideration of its radioactive properties.

Since helium occurs in considerable quantities occluded in minerals containing uranium or thorium, Rutherford and Soddy in propounding the theory of atomic disintegration suggested that helium might be one of the ultimate products of radioactive change, the gas in such minerals having gradually accumulated in the course of geological time. Rutherford made the further suggestion that the α-particle might be an atom of helium. The correctness of this suggestion was verified in 1903 by Ramsay and Soddy, who purified the gases obtained by dissolving quantities of radium bromide of the order of 50 mg. in water. After 4 days the complete spectrum of helium was obtained showing that the gas had been produced from the emanation.

The identity of the α-particle and the helium atom was established in 1908–9 as the result of some very elegant experimental work by Rutherford and his pupils. The first part of this research, carried out in collaboration with Geiger, involved the detection of single α-particles. This was effected by an adaptation of an experimental procedure due to Townsend for the production of fresh ions from an ionizing agency by collision with gaseous molecules subjected to a high difference of potential but below the sparking value. In this way the small ionization due to a single α-particle entering the gas could be magnified several thousand times, and the effect upon an electrometer thus rendered easily visible. The ballistic movements of the electrometer needle, corresponding to the arrival of each α-particle, were found to be discontinuous, but to range around an average figure in accordance with the principles of probability. Rutherford and Geiger then obtained a most interesting confirmation of their results by employing an optical method of counting the scintillations due to the impact of the α-particles upon a screen of zinc sulphide. This was a quantitative adaptation of the spinthariscope devised by Crookes a few years previously.

The second and final part of the investigation, carried out together with Royds, was a direct demonstration that accumulated α-particles, irrespective of their source, consist of helium. A quantity of radium emanation was introduced into a glass tube having walls sufficiently thin to enable α-particles to penetrate them completely, but strong enough to withstand external pressure. This tube was enclosed within an exhausted vessel for the resulting gas to be examined spectroscopically. In the course of six days the complete helium spectrum was observed.

Determinations of the rate of production of helium from radium have provided means for obtaining reliable values for one of the most important and fundamental constants in physical science, viz. Avogadro's number, or the number of molecules in one gramme-molecule of a gas. From a knowledge of the number of α-particles expelled from a known mass of radium per unit of time, Rutherford estimated the rate of production of helium at 158 cubic millimetres per gramme of radium per annum.

A more elaborate investigation by Rutherford and Boltwood in 1911 gave a value of 156 cubic millimetres of helium. From these figures the value of Avogadro's number is $6 \cdot 16 \times 10^{23}$. This result is in good agreement with calculations made from other considerations. Thus studies on the negative charges on gaseous ions associated with the names of J. J. Thomson, Townsend, C. T. R. Wilson, and later with that of Millikan have resulted in assigning a value of $6 \cdot 1 \times 10^{23}$ to the constant. A most interesting and important series of experiments started by Perrin in 1908 on the statistical behaviour of grains of gamboge suspended in water resulted in obtaining a value of the order of $6 \cdot 8 \times 10^{23}$ for Avogadro's number. These experiments on the Brownian movement have been considered to furnish some of the most convincing evidence in favour of the physical reality of molecules, and incidentally of the fundamental correctness of the kinetic theory of gases. Other evidence supporting the value of Avogadro's number has been obtained from studies of the opalescence of fluids near the critical point (Smoluchowski, 1907; Keesom, 1908–11) and also from developments arising from Rayleigh's ideas regarding the scattering action of material particles accounting for the blue colour of the sky.

Since the transformation of radium into the emanation involves the loss of an α-particle, and since the atomic weight of radium had been shown to be very nearly 226, the atomic weight of the emanation should be 222 since that of helium is 4. The value of 220 obtained from effusion experiments by Debierne in 1910 was the nearest approach to the true value. In 1911 Whytlaw-Gray and Ramsay made a direct determination of the density of a known volume of the emanation with the aid of a quartz micro-balance sensitive to about one-millionth of a milligramme. Assuming the emanation to be a monatomic gas, they concluded that its atomic weight was about 223, as the result of the average of five determinations of which the extreme values were 218 and 227. Having regard to the weight of the emanation used in these determinations being less than one-thousandth of a milligramme, the result must be regarded as highly satisfactory.

As the study of radioactive phenomena progressed the existence of a number of new elements came to be recognized, and the

problem of fitting them into the periodic table became of importance. The keys to the solution of this problem were the nature
of the α- and β-particles. The α-particle had been shown to be
a helium atom carrying two charges of positive electricity, and
the β-particle is an electron, viz. one atomic charge of negative
electricity. In 1913 suggestions which resulted in the solution
of the problem were put forward independently by Russell, by
Soddy, and by Fajans. The now well-known displacement law
is associated with the names of these investigators, and may be
expressed in the following terms. When a radio-element suffers
a loss of an α-particle by atomic disintegration, the position of
the resulting element in the periodic table is in a group two
places below it. Thus when radium, a second group element,
loses an α-particle and produces the emanation, this gas belongs
to the inert elements of the zero group. When, on the other hand,
a radioactive change involves the loss of a β-particle, the position taken up by the newly formed element involves a shift of
one place in the opposite direction. It will have been noted
that an α-particle change involves a loss of four units of atomic
mass, whereas the change in atomic mass involving the loss of a
β-particle is almost negligible. It is true that the mass of an
electron is about 1/1840 of that of a hydrogen atom, but even
this minute change of atomic weight does not arise, for after the
loss of the β-particle, the newly formed atom has its positive
charge compensated by acquiring a negative electron from
without.

One of the most interesting consequences of the study of the
chemistry of the radio-elements, and particularly of the displacement law, was the discovery of the element protoactinium
by Hahn and Meitner, and also independently by Soddy and
Cranston in 1918, who first named it ekatantalum. Since
actinium is undoubtedly a rare earth element, and belongs to
the third group, it follows that it might be derived from an
element belonging to the fifth group by the loss of an α-particle,
or, alternatively, from a second group element by the loss of a
β-particle. For various reasons the latter alternative seemed
improbable, and consequently experiments were made for a
search for an element chemically similar to tantalum in residues
from pitchblende. Eventually both sets of investigators proved

the presence of a long-lived element which accompanied the tantalic acid, emitting α-rays, and from which the production of actinium could be definitely proved.

Much experimental work was carried out on protoactinium by Hahn and Meitner in the years following the discovery of the element. It was found that this element could be obtained free from other radioactive substances without much difficulty. On the other hand, the preparation of compounds of chemically pure protoactinium was a problem of exceptional difficulty on account of its similarity to tantalum, the only important difference being that protoactinium is more basic than tantalum, just as tantalum is less acidic than niobium. Hahn and Meitner in 1921 attacked the difficult problem of the half-life period of the new element and its content in uranium minerals. In 1928, in collaboration with Walling, Hahn estimated the half-life as somewhere of the order of 20,000 or 27,000 years, by first removing all the protoactinium from a uranium salt and then after several years separating and determining the amount of the element formed from the uranium. The half-value period of protoactinium is therefore much longer than that of radium. Nevertheless, determinations of its amount in uranium minerals by Hahn and Meitner indicated that the proportion was of the order of one-fifth of that of the radium. The explanation of these somewhat anomalous figures is to be sought in the fact that only some 3 or 4 per cent of the uranium atoms disintegrate via protoactinium, the others disintegrate via the uranium-X, ionium, radium series. Since 1927 quantities of compounds of protoactinium, measured in milligrammes, were prepared by von Grosse, and also by Graue and Kading. In 1935 von Grosse determined the atomic weight from the ratio K_2PaF_7/Pa_2O_5 and obtained a value of $230 \cdot 6 \pm 0 \cdot 5$. This is definitely below that of thorium, the atomic weight of which is $232 \cdot 1$, which recalls the problem of tellurium and iodine in the periodic table.

Of the ultimate products of radioactive change, exceptional interest is attached to lead. Having regard to the universal presence of lead in uranium and thorium minerals, it occurred to Soddy and independently to Fajans that the atomic weight of lead separated from such minerals might be appreciably different from that of ordinary lead. Assuming that the lead in these minerals has originated from the atomic disintegration of uranium or thorium atoms, the atomic weight of that element should be determined by the total number of α-particles, that is, of helium atoms, each of four units of atomic mass, lost as the result of the destruction of the atoms of the uranium or thorium respectively. The successive radioactive changes from uranium to lead involve a total loss of eight α-particles, and those from thorium to lead involve altogether a loss of six α-particles. As the atomic weights of uranium and thorium are 238 and 232 respectively, it follows that the atomic weight of lead derived from the former should be 206, and that of lead produced from the latter should be 208. The atomic weight of ordinary lead is very nearly 207.

This problem was attacked by Soddy and Hyman in 1914. They extracted and purified the lead from a Ceylon thorite, and eventually converted the resulting lead chloride into silver chloride. As a control a well-purified specimen of ordinary lead chloride was converted into silver chloride under precisely similar experimental conditions. Taking the atomic weight of the ordinary lead as 207·1, Soddy and Hyman obtained a value of 208·4 for the atomic weight of their thorite lead. It should be added that this mineral was exceptionally suitable for an experiment of this kind, as the thorium content was about 55 per cent, that of uranium between 1 and 2 per cent, and the lead content about 0·4 per cent. A low ratio of uranium to thorium is of vital importance in work of this kind, because it is essential that the error due to the presence of any lead resulting from the disintegration of uranium should be reduced to a minimum, particularly as it has been estimated that the rate of atomic disintegration of uranium is between two and three times as great as that of thorium. In 1918, Fajans purified the lead from a Norwegian thorite having a still lower ratio of uranium to thorium than Soddy's mineral: the composition was 30·1 per cent of thorium, 0·45 per cent of uranium, and 0·35 per cent of lead. The atomic weight of this lead as determined by Hönigschmid was found to be 207·9.

The atomic weight of lead derived from uranium minerals, as free from thorium as possible, was determined by Richards and Lembert in 1913. Values approximating to 206·5 were obtained, although relatively marked variations were obtained as regards lead acquired from different minerals. It would seem that the lead in these minerals was not wholly derived from uranium. Hönigschmid and Horovitz in 1915 made an investigation of lead obtained from a very pure crystalline pitchblende from Morogoro in East Africa, and also from bröggerite from Norway, and obtained values for the atomic weight of 206·05 and 206·06 respectively. It should be added that in this year Merton showed that the spectra of lead with an atomic weight of 206·05 and ordinary lead were completely identical. In 1916, Richards and Wadsworth published the results of their determinations of the atomic weight of lead obtained from various uranium minerals from widely different sources. From Australian carnotite and from Norwegian cleveite values of 206·34 and of 206·08 were obtained respectively.

It will be evident how very far-reaching are the conclusions to be drawn regarding the nature of the chemical elements as enlarged by the study of radioactive phenomena. In the closing years of the nineteenth century it was universally accepted that the elements were immutable and of fixed atomic weight. Studies of the radioactive elements has shown conclusively that the atoms of these elements are not stable, but disintegrate according to well-defined principles. Still more remarkable has been the recognition of the fact that the atomic weight of an element is an average value of the *isotopes* (a name due to Soddy) of which it consists. The value of the atomic weight of ordinary lead approximating very closely to 207, and thus being an average of those derived from the disintegration of uranium and thorium atoms in the course of geological time, has received strong support from determinations of the atomic weight of the element by Hönigschmid in 1919. From three Ceylon thorianites containing both uranium and thorium in the percentages of 11·8, 20·2, and 26·8 of the former and of 68·9, 62·7, and 57·0 of the latter element, Hönigschmid obtained values of the atomic weight of the lead of 207·2, 206·9, and 206·8 respectively, figures sufficiently convincing to require no further comment.

The question whether isotopes should be described as extremely similar or as chemically identical has received considerable discussion. As regards the density of the isotopes of lead it was pointed out by Soddy in 1921 that there is a complete parallelism between the values of the atomic weight and the specific gravity. The values of the atomic volumes of ordinary lead and of lead of radioactive origin differ by less than three parts in ten thousand.

The recognition of the identity of the chemical properties of ordinary elements and of their radioactive isotopes has been turned to practical account by Paneth and his collaborators, who have introduced some of the radio-elements as indicators in analytical and other branches of chemistry. Bearing in mind that thorium C is chemically identical with bismuth, Paneth in 1918 attempted to obtain evidence regarding the existence or non-existence of a volatile bismuth hydride similar to the hydrides of antimony and arsenic. The time for the activity of thorium C to fall to one-half of its initial value is 60·5 minutes, so the element is admirably adapted for experimental work of this kind. The experiments were conducted by allowing magnesium turnings to be exposed to radiothorium, and after the decay of the thorium emanation they acquired a deposit of the disintegration products including thorium C. Hydrogen was then generated by dissolving the magnesium turnings in hydrochloric acid, and the issuing gas tested by an electroscopic arrangement for the presence of volatile radioactive substances. The results were perfectly definite: quantities of a hydride of thorium C of the order of 10^{-15} g. were shown to be present. Having found that the radioactive isotope of bismuth could yield a hydride, it followed that ordinary bismuth must do the same, and experiments on a large scale by Paneth and Winternitz verified the correctness of this view. These successful experiments were followed in 1920 by the discovery by Paneth and Nörring of a lead hydride, and of a stannic hydride by Paneth and Rabinowitsch in 1924. It is noteworthy that metals such as lead and tin which can give rise to two well-defined series of compounds, always produce hydrides corresponding with their highest valency, or in other words with their position in the periodic classification.

Another example of the use of radio-elements as indicators was the determination of the solubility of lead chromate by Hevesy and Paneth as long ago as the year 1913. Hitherto the solubility of such very sparingly soluble substances had been determined chiefly by measurements of the electrical conductivity of the saturated solutions. By using the radioactive isotope of lead, thorium B having a half-value period of 10·6 hours, accurate and reliable determinations of the solubility of lead chromate were made in a simple manner as follows. To a definite quantity of a soluble lead salt a known quantity of thorium B (as measured in electroscopic units) was added. From the lead salt now activated by thorium B the chromate was prepared, and a known volume of a saturated solution evaporated to dryness. The activity of the invisible residue on the dish was then determined electroscopically, and the concentration of the solution thus ascertained. The value found for the solubility of lead chromate, namely, 2×20^{-7} mol. per litre at room temperature, is in good agreement with that obtained by the conductivity method.

The use of radio-elements as indicators has extended far beyond investigations in pure chemistry. These elements have proved of great value in biological experiments, such as following the course of heavy metals in the organism in therapeutic research. More recently, as the result of the discovery of artificial radioactivity, it has been found possible to study certain aspects of chemical reactions which could not have been investigated otherwise. Thus Grosse and Agruss in 1935, by using a radio-isotope of bromine, were able to show that the bromide ion in sodium bromide undergoes exchange with elementary bromine in aqueous solution.

It is worthy of note that the theory of atomic disintegration, as put forward by Rutherford and Soddy in 1903, is in no way dependent on any particular theory of atomic structure. Indeed, it would be correct to say that although at that time electrons were recognized as universal constituents of atoms, and important beginnings had been made in formulating an electrical theory of matter, the original theory of atomic disintegration involved no assumptions beyond regarding the atoms of elements of high atomic weight to be unstable. Soddy has been particularly emphatic on this subject. Much interest is, however, attached to a discovery of feeble radioactivity of two elements of somewhat low atomic weight by Campbell and Wood in

1907. They showed that potassium and rubidium emit a feeble β-radiation, of the order of one-thousandth of that for corresponding quantities of uranium. Further experiments by Campbell and others have left little doubt that this feeble radioactivity is a definitely atomic property of the two elements. In 1919, Hahn and Rothenbach made some further experiments on the radioactivity of rubidium and have confirmed this conclusion. As only β-rays are emitted from each element, it follows from the displacement law that the disintegration product of potassium should be an isotope of calcium and that of rubidium should be an isotope of strontium.

The year 1913 is an important date in the history of atomic science. Previous to that time the theory of atomic structure which was generally accepted was that put forward by J. J. Thomson in 1904, who regarded the atom as consisting of a sphere of positive electrification containing a number of negative electrons distributed within it. But in 1913 several independent discoveries led to the abandonment of Thomson's original idea in favour of considering the positive electricity concentrated in a small nucleus at the centre of the atom and the electrons revolving in orbits around it. The most convincing evidence in favour of a small nucleus as opposed to a uniform sphere of positive electrification was obtained from Rutherford's experiments on the passage of α-rays through matter. The results obtained in this way were altogether inconsistent with the idea of the positive electricity being uniformly distributed throughout the atom. The model atom as conceived by Rutherford and by Bohr in 1913 may be said to show some resemblance to a micro-solar system, but it must be particularly emphasized that the size of the nucleus, which contains practically the whole of the mass of the atom, is exceedingly small in comparison with the dimensions between the nucleus and the electrons. In the Rutherford-Bohr atom it is the peripheral electrons which are chiefly concerned with the ordinary chemical properties and reactions of the element, whereas the mass and the radioactive properties of the atom are concentrated within the nucleus itself. This conception of atoms will now make it intelligible why radioactive processes, in contradistinction to ordinary chemical changes, are uninfluenced by changes in the external physical conditions. This distinction was emphasized at the outset by Rutherford and Soddy in 1903 who insisted that radioactive change was an atomic and not a molecular process. Rutherford was the first to realize that if the nucleus of an atom could be attacked by the impact of a swiftly moving α-particle it might be possible to disrupt the nucleus and thus effect the artificial destruction of the atom. After some preliminary experiments, Rutherford in 1919 succeeded in attacking the nucleus of the nitrogen atom by α-particles and observed that charged hydrogen atoms were produced.

This result was considered at the time to be due to destruction of the nuclei of the nitrogen atoms as the result of collision with the swiftly moving α-particles. It will be appreciated that in experiments of this kind it is only an exceedingly small fraction of the nitrogen atoms which suffer disruption in consequence of such collisions, as the vast majority of the α-particles pass through the nitrogen atoms unchanged, because the diameter of the nucleus is so very small in comparison with the distances between it and the electrons. The liberation of hydrogen atoms from nitrogen atoms seems to have been at first regarded as a process brought about as a result of the impact of the α-particles, in a manner analogous to the explosion of a highly unstable compound due to the action of a detonator. A few years later, especially as the result of some most interesting experiments carried out by Blackett, a different explanation was found to be necessary. These experiments consisted in following the course of the α-rays through nitrogen by photographing their path in a Wilson ionization chamber. In one set of experiments, carried out in 1929, Blackett found that out of 270,000 α-ray trajectories, eight were found to be abruptly forked. The conclusion drawn from a study of these photographic experiments was that a fork is an indication of a collision between an α-particle and the nucleus of a nitrogen atom, the α-particle being actually captured within the nucleus, thus producing a new unstable system which then decomposes into an isotope of oxygen and a hydrogen atom (proton). This result may be expressed by a nuclear chemical equation, in which the atomic weights are appended to the symbols:

$$N(14) + He(4) = N(13)He(4) + H(1).$$

Rutherford's successful attack on the nucleus of nitrogen atoms with α-particles resulting in the emission of protons was soon followed by similar results obtained with other elements of low atomic weight, such as boron, fluorine, magnesium, aluminium and phosphorus. A considerable number of investigators studied the various problems which thus arose, and the greatest interest is attached to certain results obtained by Chadwick and also independently by Irène Curie and F. Joliot since 1932. It was found that when beryllium atoms were bombarded by α-particles from polonium, protons were not emitted, but a very penetrating radiation was produced. The explanation which was given of these remarkable results was that in certain cases, such as that of beryllium, the attack on the nucleus by the α-particle resulted in the formation of a new but unstable nucleus accompanied by the emission of an uncharged hydrogen atom, called a *neutron*. The unstable atomic nucleus thus produced then disintegrated with emission of the very penetrating radiation. This subject, with which the names of I. Curie and F. Joliot are more especially identified, has been rightly named as artificial radioactivity. The attack on the nucleus of a light atom by an α-particle can take place in two entirely different ways resulting in the formation of a new *stable* atomic nucleus with the emission of a proton, or alternatively, with the formation of an *unstable* atomic nucleus together with the emission of a neutron. In the latter case the unstable nucleus undergoes atomic disintegration, just like a naturally radioactive atom does, with the emission of a positive electron, usually known as a *positron*. In the case of an attack with α-particles upon aluminium atoms, the two types of nuclear reaction may be expressed symbolically as follows:

$$Al(27) + He(4) = Si(30) + H(1)$$

(formation of a stable isotope of silicon together with the emission of a proton),

and
$$Al(27) + He(4) = P(30) + n(1)$$

(formation of a radioactive isotope of phosphorus accompanied by a neutron).

The latter reaction is then followed by atomic disintegration of the unstable phosphorus atom with formation of the stable silicon atom, having the same atomic mass accompanied with the emission of a positron.

It will be evident that these experimental results, brilliantly interesting as they undoubtedly are, have not simplified the problems of atomic structure. The recognition of atoms consisting of protons and electrons had for many years provided a basis for the explanation of numerous and varied phenomena in physics and chemistry. Since 1932, however, we have become acquainted with no fewer than four types of corpuscle which are regarded as elementary. It is permissible to regard the proton and neutron as two different states of one and the same particle. Viewed in this way, the change from proton to neutron takes place with the creation and emission of a positive electron, while the converse process, namely, the change from neutron to proton must take place with the creation and emission of a negative electron. In discussing this subject, Louis de Broglie has directed attention to one fundamental difference between the electron and the positron. The former is constantly manifesting itself in our experiments, whereas the latter appears only in altogether exceptional circumstances and has, moreover, a strong tendency to disappear when in contact with matter.

Heisenberg in 1932 enunciated some interesting views on the nature of the atomic nucleus. He considered this to consist of certain numbers of protons and neutrons. For light atoms, say up to that of calcium with an atomic weight of 40, these are present in equal numbers, and the charge on the nucleus is equal to the number of protons. As the atomic weight of an element is equal to the sum of the weights of the protons and neutrons composing the nucleus, it follows that for the light atoms the value of the atomic weight is double that of the atomic number. As we pass to the elements of higher atomic weight, the ratio of neutrons to protons exceeds unity, and when the ratio exceeds a value of 1·5, the nucleus becomes unstable and radioactive phenomena become apparent. According to Heisenberg, the expulsion of an α-particle signifies the expulsion of a helium atom consisting of two neutrons and two protons. In a radioactive change involving the loss of an α-particle the net effect would be to raise the ratio of neutrons to protons. The process of β-ray disintegration is regarded as due to the overstepping of the ratio of neutrons to protons beyond a certain limit, the expulsion of the β-particle converting a neutron into a proton, the net effect being a lowering of the ratio of neutrons to protons.

In viewing the study of radioactive phenomena in retrospect since the fundamental discoveries of Becquerel and Madame Curie, it can be seen that the earlier work involved the establishment of the theory of atomic disintegration by Rutherford and Soddy in 1903. This was of first-rate significance for chemical science, but it may be noted that many chemists at that time failed to recognize its importance. The next steps were perhaps even more far-reaching. Thus the work of Soddy and of others on the atomic weight of lead derived from radioactive minerals has given a new significance to the fundamental ideas regarding atomic weights. The recognition of isotopes, not as exceptional types identified with a very few elements, but as characteristic of the majority of them, was also of much importance. Then the displacement law of radioactive change enunciated in 1913 enabled the radio-elements to be placed within the periodic system in an ordered and regular manner. Indeed, the modern conception of atomic number is in some respects an outgrowth of the displacement law. These may be cited as some of the contributions made to chemical science from the studies in 'classical' radioactivity. What might be described as the second epoch of radioactive research can be dated from Rutherford's attack on the nucleus of the nitrogen atom in 1919 with α-particles and followed with the production of artificial radio-elements by I. Curie and F. Joliot and others. The study of artificial radioactivity appears to be attracting ever-increasing attention, and the attacks on atomic nuclei have been carried out, not only with α-particles, but with protons, deuterons, and neutrons, and results of much value, not only to pure physical science but to technical problems, may be expected.

REFERENCES

E. RUTHERFORD. *Radioactive Substances and their Radiations*. Cambridge, 1913.

E. RUTHERFORD, J. CHADWICK and C. D. ELLIS. *Radiations from Radioactive Substances*. Cambridge, 1930.

MADAME CURIE. *Radioactivité*. Paris, 1935.

J. JOLY. *Radioactivity and Geology*. London, 1909.

R. A. MILLIKAN. *Electrons*. Cambridge, 1936.

VICTOR HENRI. *Matière et Énergie*. Paris, 1933.

J. PERRIN. *Brownian Movement and Molecular Reality*. Translated by F. Soddy. London, 1910.

J. PERRIN. *Grains de Matière et de Lumière*. Première Partie: *Existence des Grains*. Deuxième Partie: *Structure des Atomes*. Troisième Partie: *Noyaux des Atomes*. Quatrième Partie: *Transmutations Provoquées*. Paris, 1935.

F. JOLIOT et IRÈNE CURIE. *Radioactivité Artificielle*. Paris, 1935.

F. PANETH. *Radio Elements as Indicators*. New York, 1928.

O. HAHN. *Applied Radiochemistry*. New York, 1936.

C. ROSENBLUM. *Chemical Reviews*, 1935, XVI, 99.

G. T. SEABORG. *Chemical Reviews*, 1940, XXVII, 199.

F. SODDY. The Conception of the Chemical Element as Enlarged by the Study of Radioactive Change. *J. Chem. Soc.* 1919, p. 1.

F. SODDY. The Part Played by Chemistry in Modern Atomic Science. *The Chemical Age*, 25 May 1935, p. 459.

Annual Reports of the Chemical Society since 1904.

Chapter V

ELEMENTS, ISOTOPES AND ATOMIC NUMBERS

Although the development of chemical science in the latter years of the nineteenth century had been powerfully stimulated by the principles laid down by Mendeleeff, it cannot be denied that chemists in the early years of the twentieth century regarded the periodic law as having to some extent outgrown its usefulness. The glamour surrounding the discovery of new elements, the properties of which had been predicted with remarkable accuracy, had faded, and instead there was a development of interest in some of the limitations of the periodic law. It was gradually realized that the periodic law must be to some extent an approximation, and that there must be some more fundamental principle behind it still awaiting discovery. Eventually it was found that the atomic weight of an element could no longer be regarded as its most fundamental characteristic, and that elements might consist of atoms of more than one atomic mass, the atomic weight being an average value of those of the isotopes. The study of radioactive phenomena, of the X-ray spectra of the elements, and of analysis by rays of positive electricity, were the chief means by which some of the most fundamental concepts in chemistry received such striking and profound alteration.

It may be convenient to consider spectroscopy first. The foundations of spectrum analysis laid by Bunsen and Kirchhoff in 1860 soon resulted in the discovery of several new elements, e.g. thallium by Crookes in 1861, and indium by Reich and Richter in 1863. This kind of work was to a large extent empirical. There remained the difficult problem of the origin of spectra. In 1885 an important law governing the spacing of the lines in the spectrum of hydrogen was discovered by Balmer, and further elaborated by Rydberg (1890), which amounted to representing each line in the hydrogen spectrum by the difference between two terms of the form R/m^2, where R is known as Rydberg's constant and m is an integer. This law was formulated as a result of studies on the lines of the *visible* region of the hydrogen spectrum. Later it was found that the same type of expression was applicable to the spectral lines on both sides of the visible region, namely, to the Paschen series in the infrared region discovered in 1908, and to the series in the extreme ultra-violet discovered by Lyman in 1916. It was also found that the spacing of the lines in the spectra of other elements could be represented in the same sort of way. These results, although interesting and valuable, were not properly understood until the year 1913 when Bohr correlated the Rydberg constant with the charge and mass of an electron, Planck's constant, and the velocity of light. In the same year Moseley showed that the X-ray spectra of the elements could be brought into line with the Balmer-Rydberg law.

In the study of the rare earths, spectroscopic methods have been of considerable importance. Emission spectra may be produced by vaporizing substances either by thermal or electrical methods. Thus we may have flame, arc, or spark spectra. In 1881, Crookes began a series of studies on what he termed phosphorescence spectra. This subject was closely connected with the early work on the cathode-rays. His method consisted in placing the substance under investigation in a highly exhausted glass bulb and subjecting it to a powerful electric discharge. Under these conditions many substances were found to become phosphorescent, and spectroscopic analysis of the light at first proved of much value in following the course of the fractionations. Later, however, difficulties arose in the interpretation of some of the results. Frequently Crookes and others observed that continued purification of an earth by some process of fractional crystallization of the salts would result in a concomitant intensification of the phosphorescent spectrum. It was naturally supposed that the purification process was effecting the desired object of concentrating the element causing the phosphorescence. Disappointment, however, soon followed when further purification caused the phosphorescence to disappear. In work of this kind a purely chemical control of the fractionation was obtained by determinations of the mean equivalent weight of the fractions. Eventually the puzzling results observed with the phosphorescent spectra were explained. It was found that *pure* substances do not phosphoresce. The property of producing these phosphorescent spectra appears to be essentially one of substances in solid solution. A familiar example of this kind of phenomenon is to be found in the Welsbach incandescent gas mantle which gives a maximum degree of illumination when the mixture of oxides consists of 1 per cent of cerium dioxide with 99 per cent of thorium oxide. Other proportions give rise to a marked lowering of the illuminating power. It is not difficult to understand how Crookes and others were led to erroneous conclusions regarding the presence of considerable numbers of elements in the rare earths. These somewhat fantastic ideas were strongly opposed by Lecoq de Boisbaudran (1885–90) who employed different spectroscopic methods from those of Crookes. One of the methods used by Lecoq de Boisbaudran consisted in observing the spectra produced by passing sparks between a platinum electrode and the surface of the solution under investigation.

The last of the rare earths to be discovered by purely chemical methods aided by 'classical' spectroscopy was lutecium. Urbain in 1907 fractionated ordinary ytterbium nitrate and obtained end fractions with metals having atomic weights of 169·9 and 173·8 respectively. The latter figure was assigned to the new element lutecium. It should be added that this element was discovered independently by Auer von Welsbach who named it cassiopeium. In all this work the assistance given by spectrum analysis was of a strictly analytical character, and wholly unconnected with any physical views regarding the origin or nature of spectra.

The introduction of X-ray spectroscopy by Moseley in 1913 marks the beginning of new and fundamental conceptions regarding the elements. It seems not unlikely that Moseley was attracted to this study as a result of the success of the Braggs's work on the diffraction of X-rays by crystals. Moseley made a careful investigation of the spectra produced when different substances were used as the anticathodes of X-ray bulbs. He found that the characteristic spectra thus produced were of a remarkably simple character, and consisted of two main lines the positions of which were related in a regular manner to the order of the places occupied by the elements in the periodic system when numbered consecutively. It was in this way that the *atomic numbers* of the elements were first clearly defined. Moseley's general method was to plot the square root of the frequency of the rays against the atomic numbers of the elements when simple and approximately linear relations were obtained. For the lighter elements between aluminium and silver, the K-series of characteristic X-rays was examined, and for the heavy elements between zirconium and gold, the L-series. The wave-lengths of the K-series of X-rays range from 8·4 to 0·56, and those of the L-series range from 6·09 to 1·29 Ångstrom units. Moseley's fundamental law, published in 1913, effected what was little short of a revolution in the conception of the elements. In the first place it became immediately apparent that the atomic number, and not the atomic weight, was the fundamental constant which governed the nature of an element, and in particular its place in the periodic classification. Thus the inversion of the order in the cases of argon and potassium, cobalt and nickel, and tellurium and iodine, based upon atomic weights, was rectified. Secondly, it was shown that the number of elements between hydrogen and uranium (counting those two) was 92. Soddy in 1914 remarked that 'what amounts to a veritable roll-call of the elements has been made by this method'. Certain elements were missing, and some have, as will be seen presently, been subsequently discovered. Of particular interest was Moseley's demonstration that only eighteen elements exist between barium with an atomic number of 56 and tantalum of atomic number 73 (counting those two). This determines the number of rare earths including lanthanum of atomic number 57 as fifteen, and is of very great importance, because previous to Moseley's generalization it was impossible to deny that a larger number of rare earths might not exist. At the present time element number 61 seems to be the only doubtful rare earth, and the name of illinium has been given by more than one chemist who has claimed, perhaps prematurely, to have discovered it.

When Moseley's principle was first published, attention was drawn to six elements, namely, those of atomic numbers 43, 61, 72, 75, 85, and 87, which were then undiscovered. Of these elements, numbers 85 and 87 must belong to the halogens and the alkali metals respectively, and are likely to be radioactive and of short half-value period. Their discovery is therefore likely to present considerable difficulty. Element 91 does not appear to have been particularly noted as missing, but its discovery by Hahn and Meitner in 1918, who named it proto-actinium, and independently by Soddy and Cranston who named it ekatantalum, has been mentioned in Chapter IV.

Exceptional interest surrounds the discovery of element 72. As long ago as the year 1911, Urbain claimed the existence of a rare earth element to which he assigned the name of celtium. In 1922, Dauvillier examined the X-ray spectra of certain rare earths and obtained some evidence giving possible indications of the reality of Urbain's new element. The problem was solved by Hevesy and Coster in 1923. They pointed out that the complete interpretation of the X-ray spectra of the elements must be considered alongside Bohr's theory of atomic structure (1921–2). In particular, Bohr's electronic arrangement postulates that in passing from one element to the next, the new electron usually goes to the outermost shell; but exceptions occasionally arise, and the new electron may have a preference for one of the inner shells. Proceeding in this way it became evident that lutecium with an atomic number of 71 must be the last of the rare earths, and element 72 according to Hevesy and Coster could not belong to this class of substances, but must be assigned to the fourth group in the periodic table, and have properties akin to those of zirconium.

The conclusion that element 72 must be quadrivalent and similar to zirconium in properties is perhaps one of the most interesting consequences of Bohr's theory of atomic structure. The beginnings of this theory can be traced as far back as the year 1912, when van den Broek ventured the suggestion that the atomic number of an element in the periodic system is equal to the number of electrons in its atom. In the following year Bohr pointed out that Rutherford's ideas relating to atomic structure could not be reconciled with classical electrodynamics, but required the quantum theory. It was not until 1921 that Bohr embarked upon the problem of the actual distribution of the electrons in the atomic shells and the characterization of the electronic orbits. This was done very largely as a result of his studies on spectral lines. A more purely chemical approach to this problem, which was a decided advance upon the older ideas of Lewis and Langmuir, was put forward by Bury in 1921, who actually predicted the discovery and electronic configuration of the element 72. He pointed out that the atom of lutecium (atomic number 71) would have a structure of 2, 8, 18, 32, 8, 3 electrons in the successive shells around the central nucleus, and that element 72 between lutecium and tantalum would have a structure of 2, 8, 18, 32, 8, 4 electrons in successive shells, and would therefore be quadrivalent like zirconium.

The correctness of this reasoning was established by the experimental work of Hevesy and Coster in 1923. They extracted samples of Scandinavian zircon with acids to remove soluble constituents and examined the residues by X-ray spectroscopy. The presence of the missing element was clearly indicated, and by a laborious series of fractional crystallizations of the potassium zircono-fluorides it was found that the separation could be effected, the new element, to which the discoverers gave the name of hafnium, being obtained from the more soluble fractions of the complex fluorides. Other salts such as the phosphates were also used for separating hafnium from zirconium. Actually the two elements resemble each other very closely in chemical properties, and in most zircons the hafnium content, as expressed in terms of zirconium, is of the order of 2 per cent. This subject has an important bearing upon the atomic weight of zirconium. A determination by Marignac in 1860 of the ratio K_2ZrF_6/ZrO_2 gave a value of 91·54. Hevesy and Coster considered that the content of hafnium oxide in the zirconium oxide finally weighed would have been about 0·5 per cent, which would bring Marignac's figure for the atomic weight of zirconium down to 91·22. In 1924, Hönigschmid and Zintl redetermined the atomic weights of both elements from the ratios of the bromides to silver and obtained values of 91·22 and 178·6 respectively.

The elements having the atomic numbers 43 and 75 remained undiscovered until the year 1925, although in one version of the periodic table Mendeleeff had left spaces for elements having atomic weights of about 99 and 188, to which the provisional names of eka-manganese and dwi-manganese were assigned. In 1925, Noddack, Tacke, and Berg announced the discovery of the elements having these two atomic numbers as the result of X-ray spectroscopy applied to certain minerals. The element 43 was named masurium, and the name of rhenium was assigned to element 75. Since that time little progress has been made in the study of masurium, but a considerable amount of work has been done on the chemistry of rhenium, chiefly by W. and I. Noddack. The atomic weight of the metal was determined from the ratio $AgReO_4/AgBr$ by Hönigschmid and Sachtleben (1930), who obtained a value of 186·3.

In concluding this brief account of X-ray spectroscopy, it should be remembered that Moseley in 1913 assumed that the atomic number of an element represented not only its ordinal number in the periodic system, but also the magnitude of the nuclear charge. This latter assumption was shown to be correct by Chadwick in 1920 by very exact experiments on the scattering of α-particles by metals. The values thus obtained for the atomic nuclei of platinum, silver, and copper were 77·4, 46·3, and 29·3 respectively, and the corresponding atomic numbers are 78, 47, and 29. Thus the atomic number gives a direct measure of the number of free electrons in an atom, that is, of the number of electrons in its extranuclear structure. The arrangement of these extranuclear electrons is concerned with the determination of the fundamental chemical properties of the element.

It has been seen that the conception of atomic numbers has been the means of regularizing the rare earths as a special subgroup in the periodic table. The remarkable similarity in the properties of these elements has been explained as being due to entry of additional electrons, owing to increase in atomic number, into one of the *inner* layers of the electron shells, not into the outer layers. In the groups of the periodic table there is a general increase of basic or electropositive properties with increase of atomic weight. Thus in the case of potassium, rubidium, and caesium, it is well known that caesium is the most strongly basic, presumably because the valency electron of caesium is readily detached, as it is far removed from the nucleus and so heavily screened by the inner layers of electrons. With the rare earths, however, the conditions are altogether different. The basicity of these lanthanides, as they are sometimes termed, diminishes with increase of atomic number from lanthanum (57) to lutecium (71), and this is considered to be due to increase in the charge on the nucleus with increase in atomic number. This change in basic properties with increase in atomic number is correlated with the atomic volume. In 1869, Lothar Meyer constructed his atomic volume curve, which shows in a striking manner the increase in atomic volume of the alkali metals with increase in atomic weight. But with the rare earths the atomic volume actually *decreases* as we pass from lanthanum to lutecium, and the diminution in basic properties is explained by the electrons being more firmly held. This subject has received much attention from V. M. Goldschmidt since 1927 on the crystallographic side, who has discussed the *lanthanide contraction* in terms of the apparent ionic radii, and has established principles relating to crystal structure and the nature of the atoms taking part in it.

After some preliminary experiments on rays of positive electricity, J. J. Thomson in 1911 devised a method of analysis by which results of far-reaching importance for atomic science have been obtained. The method consisted in allowing the rays to pass through a very narrow tube, and subjecting them to powerful electrostatic and magnetic fields. When these directing fields are not applied, the rays impinge on a fixed point, but when they are deflected by the action of the fields they trace out a parabola which is registered on a photographic plate, because they are moving with different velocities. By making appropriate measurements of the curves, information relating to the gases in the exhausted vessel, and in particular trustworthy information regarding the values of their atomic masses, could be obtained. Of the earlier results obtained by the 'parabola' method special interest is attached to the demonstration by J. J. Thomson in 1912 of the existence of an isotope of neon having an atomic mass of 22 as compared with that of ordinary neon which is 20. This was at the time when much attention was being given by Soddy and others to the possibility of variations in the atomic weight of lead according to its source as a disintegration product of uranium or of thorium. As is well known this conclusion was brilliantly verified in the case of this particular element, but the question was naturally raised as to whether elements not of radioactive origin would show evidence of the existence of isotopes. The problem of neon received much careful experimental investigation by Aston who carried out many laborious attempts to separate the isotope of atomic mass 22 by fractional diffusion and by fractional absorption with cooled charcoal, but without any clear positive results. Aston realized that the 'parabola' method was not sufficiently sensitive to solve some of the problems connected with the atomic masses of isotopes, particularly as J. J. Thomson had found in 1918 that chlorine gave a single parabola corresponding to its chemical atomic weight. In 1919, Aston constructed an improved form of apparatus in which particles of different velocities, but having the same ratio of charge to mass, were brought to a focus upon the photographic plate, thus producing an effect resembling a spectrum. This apparatus, which has since become known as the mass spectrograph, has enabled isotopes to be characterized with greater precision than was possible with the original apparatus. In 1920, Aston investigated chlorine with the mass spectrograph and showed conclusively that the gas consisted of at least two isotopes having atomic masses of 35 and 37 respectively, and mixed in proportions such that the chemical atomic weight is 35·46.

It has been well established that the atomic weights of elements which are non-isotopic approximate to integral numbers, whereas those of elements with a definite figure after the decimal point have such values as a result of the very thorough mixing of the isotopes. The whole problem of the constancy of atomic weights thus assumed a highly interesting aspect. Before the discovery of isotopes, the skill and energy of workers on atomic weights had been directed wholly to the preparation of substances of extreme purity for the determinations, but without regard to the *origin* of the substances. After the work of Soddy and Hyman on the atomic weight of lead derived from a Ceylon thorite in 1914 and that of Fajans in 1918 on lead obtained from other thorium minerals, attention was directed to studies on atomic weights of other elements, not of radioactive origin, which were derived from widely different sources and were most unlikely to have been admixed. In almost every case constancy in the values of the atomic weights was realized, or when there was any difference it was so small as to indicate the possibility of being due to experimental error. Thus in a most careful redetermination of the atomic weight of boron carried out in 1925 by Briscoe and Robinson, in which the ratio of boron trichloride to silver was measured, it was found that the atomic weight of the elements derived from minerals obtained from Europe and Asia Minor was 10·82, while a value of 10·84 was obtained from sources in North America. This very small difference cannot be considered very convincing, and similar results showing still smaller differences have been found for the atomic weights of other elements obtained from widely different sources.

An interesting and important aspect of atomic weight research is the correlation of determinations of *chemical* atomic weights, which in the case of isotopic elements are really the weighted mean of the atomic masses of the isotopes, with direct determinations of the atomic weights by the measurements made with the mass spectrograph. As the result of various refinements, particularly of p   ometry, to the mass spectrograph, Aston has been able to obtain atomic weights as accurately as can be done by equivalent weight determinations. The atomic weight of antimony may be quoted by way of example. Until the year 1921, the value of the atomic weight was regularly given as 120·2, but a careful determination by Willard and McAlpine from the ratio of antimony bromide to silver gave a value of 121·77. In 1924 figures of 121·75 were obtained by Weatherill and of 121·76 by Hönigschmid, Zintl and Linhard. In 1922, Aston examined the mass spectrum of antimony trimethyl and found two lines corresponding to isotopes having masses of 121 and 123. He considered these results to agree well with those of Willard and McAlpine, and to definitely rule out the older value for the atomic weight. It may be noted that, although the masses of isotopes approximate fairly closely to whole numbers, there are small but definite deviations, and Aston found it necessary to introduce a correction, which he termed the packing fraction, to express the ratio of the difference between the isotopic mass and the nearest whole number to the isotopic mass. In 1931, Aston made some redeterminations of certain atomic weights with the mass spectrograph, and in the case of antimony, applying the necessary correction for the packing fraction, he obtained a value of 121·79 for the atomic weight. Aston has remarked that 'were the vast mass of chemical data lost, the Table of Atomic Weights could be reconstructed to-day, entirely from mass spectrum evidence, as complete and, with the possible single exception of the element copper, as accurate'.

The atomic weight of hydrogen had for many years been accepted as close to 1·0078 when expressed on the oxygen standard, and from the standpoint of positive-ray analysis this was considered to show a satisfactory deviation from the whole-number rule, as Aston in 1920 obtained a closely similar figure with the mass spectrograph. It was therefore concluded that hydrogen was a non-isotopic element, since the presence of an isotope would imply an atomic mass of 2. For a number of years this conclusion received general acceptance, as the existence of an isotope of double the mass of ordinary hydrogen was regarded as highly improbable. In 1932, however, Urey, Brickwedde, and Murphy subjected liquid hydrogen to an elaborate process of fractionation, and succeeded in showing the presence of an isotope of mass 2. Later this substance, first known as heavy hydrogen and afterwards as deuterium, was prepared in larger quantities by electrolytic methods of concentration, and has been the subject of numerous investigations.

Experiments on the separation of isotopes have been made by numerous methods, and considering the difficulties which necessarily attend such attempts, it is scarcely surprising that little success has been achieved. Except in the case of hydrogen, where there is an extreme difference in the masses of the two isotopes, the problems with other elements apparently must continue to offer formidable difficulties. Between 1920 and 1922, Brönsted and Hevesy obtained a partial separation of two of the isotopes of mercury by fractional evaporation of large quantities of the metal under very low pressures. The distillate was found to be richer and the residue poorer in the lighter isotope than the original liquid. Similar results were also realized by fractional effusion of the vapour. The proof of these partial separations was obtained by determinations of the density. In 1922, Harkins and Mulliken confirmed the results of Brönsted and Hevesy regarding the partial separation of the mercury isotopes. In 1923 a revision of the atomic weight of mercury was carried out by Hönigschmid and Birkenbach. The purified metal was converted into the chloride or bromide and the resulting halide reduced by hydrazine. The ammonium chloride or bromide resulting from this operation was then titrated nephelometrically with silver nitrate. The result obtained was 200·61. The method was then applied to purified mercury which had been subjected to fractional evaporation at a low temperature in a high vacuum. The atomic weights of the metal in the lightest and heaviest fractions were found to be 200·564 and 200·632 respectively. As the error in these determinations did not exceed $\pm 0\cdot007$ unit there is no doubt regarding the genuineness of the partial separations of the isotopes. Another interesting example of experimental work on problems of this kind is to be seen in the work of Harkins and Hayes in 1921 on the application of diffusion methods to hydrogen chloride to obtain partial separation of the isotopes of chlorine. They obtained differences of the order of one part in one thousand in the atomic weight of the element in the fractions.

Shortly after it became generally recognized that it was more fundamental to characterize elements by their atomic numbers as distinct from their atomic weights, expression was given to this by introducing the atomic numbers into the periodic table. Thus in the annexed table each element is characterized by both constants, and indeed this should always be done to avoid any ambiguity regarding isotopes.

In studying the artificial disintegration of elements, the use of both atomic numbers and atomic weights has been general, and the importance of this can be seen when it is borne in mind that irradiation of aluminium atoms by α-rays can result in two different modes of transmutation, eventually leading to the same stable isotope of silicon $^{30}_{14}\text{Si}$. The transmutation may take place *directly* according to the nuclear equation

$$^{27}_{13}\text{Al} + ^{4}_{2}\text{He} = ^{30}_{14}\text{Si} + ^{1}_{1}\text{H},$$

with the emission of a proton; or *indirectly* with the formation of a radioactive isotope of phosphorus and the liberation of a neutron; followed by atomic disintegration of the radio-phosphorus, $^{30}_{15}\text{P}$ into silicon with the emission of a positron, thus:

$$^{27}_{13}\text{Al} + ^{4}_{2}\text{He} = ^{30}_{15}\text{P} + ^{1}_{0}n, \quad \text{and} \quad ^{30}_{15}\text{P} = ^{30}_{14}\text{Si} + e^{+}.$$

Periodic Classification of the Elements

Group \ Period	I	II	III	IV	V	VI	VII	Transitional			VIII
1	1 H 1·0078										2 He 4·002
2	3 Li 6·940	4 Be 9·02	5 B 10·82	6 C 12·01	7 N 14·008	8 O 16·0000	9 F 19·00				10 Ne 20·183
3	11 Na 22·997	12 Mg 24·32	13 Al 26·97	14 Si 28·06	15 P 31·02	16 S 32·06	17 Cl 35·457				18 A 39·944
4	19 K 39·096	20 Ca 40·08	21 Sc 45·10	22 Ti 47·90	23 V 50·95	24 Cr 52·01	25 Mn 54·93	26 Fe 55·84	27 Co 58·94	28 Ni 58·69	
	29 Cu 63·57	30 Zn 65·38	31 Ga 69·72	32 Ge 72·60	33 As 74·91	34 Se 78·96	35 Br 79·916				36 Kr 83·7
5	37 Rb 85·48	38 Sr 87·63	39 Y 88·92	40 Zr 91·22	41 Nb 92·91	42 Mo 96·0	43 Ma —	44 Ru 101·7	45 Rh 102·91	46 Pd 106·7	
	47 Ag 107·880	48 Cd 112·41	49 In 114·76	50 Sn 118·70	51 Sb 121·76	52 Te 127·61	53 I 126·92				54 X 131·3
6	55 Cs 132·91	56 Ba 137·36	57 La 138·92	72 Hf 178·6	73 Ta 180·88	74 W 184·0	75 Re 186·31	76 Os 191·5	77 Ir 193·1	78 Pt 195·23	
	79 Au 197·2	80 Hg 200·61	81 Tl 204·39	82 Pb 207·21	83 Bi 209·00	84 Po 210	85 —				86 Rn 222
7	87 —	88 Ra 226·05	89 Ac 227	90 Th 232·12	91 Pa 231	92 U 238·07					

The Rare Earths (to be placed between 57 La and 72 Hf)

58 Ce 140·13	59 Pr 140·92	60 Nd 144·27	61 Il —	62 Sm 150·43	63 Eu 152·0	64 Gd 156·9
65 Tb 159·2	66 Dy 162·46	67 Ho 163·5	68 Er 167·64	69 Tm 169·4	70 Yb 173·04	71 Lu 175·0

It will be evident that an element of atomic weight 30 might *a priori* be an isotope of either silicon or of phosphorus, but if labelled with both atomic number and atomic weight it is completely characterized. It is no less important to remember that an element of atomic number 82 includes at least two isotopes of lead, but these can be distinguished by the symbols $^{206}_{82}Pb$ and $^{208}_{82}Pb$, according as the particular isotope is the product of radioactive change of uranium or of thorium.

The development of theories regarding the constitution of atoms has arisen as the result of more than one line of investigation, but there would appear to be little doubt that the existence of electrons in all kinds of matter was the central source of inspiration. Even Lord Kelvin, who was dissatisfied with the theory of atomic disintegration and considered that radioactive phenomena might be explained on the basis of the absorption of radiant energy from without, studied various atomic models from the standpoint of radioactivity, but his views remained undeveloped and were of no interest from the chemical point of view.

In studying the development of views on atomic structure, particularly as regards their bearing upon the more fundamental aspects of valency and theories of chemical combination, attention must first be directed to atoms as understood by J. J. Thomson in 1904. The electrons were then regarded as embedded within a sphere of positive electricity, and arranged in concentric spherical shells around the centre of the atom, the number of shells being determined by the total number of electrons present. The problem was much simplified by considering the forces as acting in one plane in which the shells were regarded as rings. It was shown theoretically that stable arrangements would be obtained with a single ring having not more than five electrons. On adding a sixth it goes to the centre, and on adding still more electrons the ring becomes stable up to a maximum of eight, after which inner rings are formed. It was pointed out by J. J. Thomson that considerations of this kind seemed to indicate a means of explaining the periodic system of the elements. In 1907 he showed that these views on atomic structure provided a basis for discussing problems of valency on an electrochemical basis, similar in all essential respects to those enunciated by Abegg in 1904 regarding the sum of the normal and contravalencies of an element as equal to eight. This subject was further developed by J. J. Thomson in 1914, who pointed out a very fundamental distinction between polar and non-polar molecules, and remarked that 'an atom may exert an electropositive valency equal to the number of mobile corpuscles in the atom, or an electronegative valency equal to the difference between eight and this number'.

In 1916, Kossel and Lewis independently published theories of valency discussed in terms of the electronic theory of atomic structure. The ideas of Kossel were developed chiefly from Abegg's theory of normal and contravalency and from van den Broek's identification of the atomic number of an element in the periodic system as being numerically equal to the number of electrons in its atom. Much importance was attached to the atomic structure of the inert gases, all of which, except helium, were considered to contain a closed outer ring of eight electrons. Thus the atoms of helium, neon, and argon would consist of structures of 2; 2, 8; and 2, 8, 8 electrons respectively corresponding to the atomic numbers of 2, 10 and 18. It was noted that an inert gas in the periodic system was always preceded by a halogen and followed by an alkali metal. The strong tendency of the halogens to give rise to negative ions, and of the alkali metals to produce positive ions, could be expressed by the outer ring of electrons in the chlorine atom requiring one more electron to complete its complement of eight, and a corresponding process of removal of one electron from the potassium atom to effect the same result. The formation of potassium chloride was thus expressed in terms of a simple transfer of electrons. Most of Kossel's theory was concerned with ionizable compounds, and he extended it to include the consideration of complex ions. Throughout the discussion of his theory, Kossel avoided the use of any model atom, except in so far as he made some references to the ideas of Bohr.

The theory of valency put forward by Lewis had many points in common with that of Kossel, but at the outset Lewis made use of an atomic model. This was the celebrated cubical atom in which the electrons were regarded as situated at the corners of the cube. Thus the atom of lithium would have an electron at one corner of the cube, the beryllium atom would have two corners each occupied by an electron, and so on until neon was reached with all the corners occupied by electrons. The inert gas structure having a stable arrangement of eight electrons was thus realized. Lewis stressed the important distinction between polar and non-polar compounds and referred to the important contributions to the theory of chemical combination made by J. J. Thomson in 1914. In contrast to Kossel who was concerned with polar compounds, Lewis devoted particular attention to non-polar compounds. He introduced the conception of the sharing of electrons to denote a non-polar valency, and expressed this by the use of a colon or two dots to represent the two electrons which act as the connecting links between the two atoms. Thus water in Lewis's notation was expressed by the formula $H:\overset{..}{O}:H$. The ideas of Lewis were extended by Langmuir in 1920, particularly as regards the building up of stable groups of electrons, usually octets, who stressed the distinction between electrovalency in which this process is effected by the transference of electrons, and covalency which arises in consequence of the sharing of electrons between adjacent atoms.

In 1923 important contributions to the study of valency were made independently by Lowry and by Sidgwick. Lowry emphasized the uniqueness of hydrogen in giving rise to covalent and electrovalent compounds, and considered that it was not always easy to draw a distinction between the two kinds of valency. Thus anhydrous hydrogen chloride, which so long ago as 1865 was shown by Gore to be devoid of acid properties, and in 1923 was shown by Hantzsch to resemble the alkyl chlorides as regards similarity of the absorption spectra, is evidently a covalent compound. Its structure would therefore be represented by the formula $:\overset{..}{Cl}:H$. Such a formula would clearly be inappropriate for hydrochloric acid. Lowry regarded the essential distinction between a hydrogen ion (or more correctly a hydroxonium ion) and a hydrogen atom covalently linked as determined by the mobility of the former as contrasted with the attachment of the latter not merely to a particular atom, but 'to a particular region in the structure of the atom, for example by the sharing of a particular pair of electrons'.

In the study of co-ordination compounds, electronic theories of valency have resulted in a much clearer understanding of the nature of these substances than was possible according to the useful but necessarily empirical views of Werner. Thus in the formation of ammonium chloride from ammonia Lowry adverted to the nitrogen atom as having a 'lone pair' of electrons which could be made to function when the element was exerting its maximum valency. Considerations of this kind led to the recognition of a third type of linking which Lowry termed a semi-polar double bond, but the term of co-ordinate link introduced by Sidgwick has been found more suitable. The outstanding contribution made by Sidgwick to the whole subject was his recognition of the importance of the Bohr atom as a basis for any logical and consistent theory of valency. The cubical atomic model of Lewis and Langmuir was admittedly an over-simplification, as it represented the electrons in fixed positions, whereas the conceptions of Rutherford and Bohr necessitated revolutions of the electrons in circular or more generally in elliptical orbits around the central nucleus. Nevertheless, although unsound from the standpoint of physics, the cubical atom was a helpful foundation for the development of the ideas of transference and of sharing of electrons. It should be remembered that the improvement upon Langmuir's ideas put forward by Bury in 1921 on strictly chemical considerations resulted in assigning the same electronic structures to the inert gases as Bohr did at the same time; namely, for helium 2, for neon 2, 8, for argon 2, 8, 8, for krypton 2, 8, 18, 8, for xenon 2, 8, 18, 18, 8, and for niton 2, 8, 18, 32, 18, 8, in successive shells.

A fundamental difference between the conceptions of Bohr and of Bury as contrasted with those of Lewis and Langmuir is concerned with the size of the successive groups of electrons which lead to structures of maximum stability. In the atomic structures of Bohr and of Bury the larger groups are not on the outside but in the middle, and the outermost ring (except in the case of helium) always consists of eight electrons. Sidgwick has described the formation of an ammonium ion from a neutral nitrogen atom by pointing out that this atom has five valency electrons. On combining with three hydrogen atoms its octet is completed and the molecule is in one sense saturated. A nitrogen atom which has expelled an electron, and thus acquired a positive charge, has four valency electrons. It can therefore take up four hydrogen atoms (or four alkyl groups) to form an ammonium ion. The similarity between an ammonium ion and a molecule containing a single carbon atom is therefore well marked, as may be seen in studies on the optical isomerism of these substances. The contrast in chemical behaviour between ammonium chloride, a strong electrolyte, and methyl chloride, in which the chlorine atom is attached to the carbon atom by a covalent linkage, is expressed in a simple and satisfactory manner by the electronic formulae,

$$\left[\begin{array}{c} H \\ H : \overset{..}{\underset{..}{N}} : H \\ H \end{array} \right]^{+} Cl^{-} \quad \text{and} \quad H : \overset{..}{\underset{..}{C}} : Cl.$$

In ammonium salts the co-ordination number as defined by Werner is four, but other co-ordination numbers are known, particularly the number six which is very common. As an example of Sidgwick's method of discussing this subject the derivatives of quadrivalent platinum may be quoted by way of illustration. In potassium platinichloride, $\overset{++}{K_2}[\overset{--}{PtCl_6}]$, the anion is bivalent. Replacement of the halogen atoms by molecules of ammonia or water results in successive diminution of the negative charge, until the compound, $Pt(NH_3)_2Cl_4$, a non-electrolyte, is formed. Further replacement of the chlorine atoms by ammonia molecules results in the group acquiring a positive charge, until the compound $[\overset{++++}{Pt(NH_3)_6}]\overset{----}{Cl_4}$ with a quadrivalent cation is obtained.

There is no doubt that electronic theories of valency have attracted much interest in many and varied directions, and it will be interesting to follow their future developments. During the last decade there has been a marked tendency on the part of physicists to depart from the original views regarding electrons as material or subatomic substances, and to consider them as being of less importance than the undulations with which they are associated, and to which may be assigned their general behaviour. The development of these views has been much stimulated by the introduction of wave mechanics by

Louis de Broglie and others. Indications are not wanting that these newer theoretical developments have influenced the views of many chemists regarding the structure of molecules. One aspect of this may be seen in the modern concept of resonance, largely associated with the name of Pauling, which was introduced about 1931. The theory of resonance attempts to deal with the problem of the constitution of molecules in which more than one kind of linking is possible, and is at the same time strictly consistent with the chemical behaviour of the compound. Thus the molecule of benzene would be regarded as a resonance hybrid of the structures represented by the extreme limits of the Kekulé formula and of the formula suggested by Dewar, in which the para positions are linked by a bond. More than one writer has emphasized a fundamental distinction between the concept of resonance and that of tautomerism. The dynamic conception of the molecule of benzene, as understood by Kekulé, was essentially a mechanical one, the double bonds being supposed to oscillate from one position to another, whereas according to the theory of resonance no single formula can really represent the molecule completely. It must, however, be added that Pauling himself has stated that 'there is no sharp distinction which can be made between tautomerism and resonance'. Pauling has made exhaustive studies with the object of ascertaining the nature of chemical bonds, and one of his general conclusions seems to indicate that the distinction between electrovalent and covalent linkages is capable of precise and accurate definition so long as clear-cut cases are considered. In other cases, however, some kind of transition between the different kinds of linking must be admitted.

From the standpoint of pure chemistry some of these problems may lead back to the old question as to how much information is to be expected from chemical formulae. It will be recalled that as long ago as 1856 Gerhardt raised the question of using more than one formula in order to express different aspects of the chemical behaviour of substances. Although much has been done since that time in the way of improving formulae, it would seem that there have been, and apparently must continue to be, two different schools of thought regarding the sense of values to be attached to them. Some chemists, interested perhaps in the more experimental side of chemistry, have concentrated attention upon the necessity of formulae presenting accurate pictures of the substances which they represent, without entering very much into the more difficult questions of the nature of the forces which hold the atoms and molecules together. Others, whose interests have centred chiefly in theoretical questions, have been very much concerned with the importance of seeing that all formulae shall be consistent with certain general principles, such as electron distribution and energy states, even though alternative formulae may be no less successful in giving expression to the general properties and reactions of the substances. In recent years this latter point of view has tended to predominate, as may be seen in the considerable amount of attention which has been given towards establishing flawless formulae for simple molecules, such as those of carbon monoxide, nitrous oxide, and diazomethane, the reactions of which are as readily intelligible on the older formulae as on those which have been advocated at the present time.

REFERENCES

J. J. THOMSON. *The Corpuscular Theory of Matter*. London, 1907.

J. J. THOMSON. *Rays of Positive Electricity and their Application to Chemical Analysis*. Second edition. London, 1921.

J. J. THOMSON. *The Electron in Chemistry*. Five Lectures to the Franklin Institute, 1923.

F. W. ASTON. *Mass Spectra and Isotopes*. London, 1942.

R. W. WOOD. *Physical Optics*. Third edition. New York, 1936.

NIELS BOHR. *The Theory of Spectra and Atomic Constitution*. Second edition. Cambridge, 1924.

G. VON HEVESY. *Chemical Analysis by X-rays and its Applications*. New York, 1932.

W. KOSSEL. Ueber Molekülbildung als Frage des Atombaus. *Ann. Phys., Lpz.*, 1916, XLIX, 229.

G. N. LEWIS. The Atom and the Molecule. *J. Amer. Chem. Soc.* 1916, XXXVIII, 762.

C. R. BURY. Langmuir's Theory of the Arrangement of Electrons in Atoms and Molecules. *J. Amer. Chem. Soc.* 1921, XLIII, 1602.

T. M. LOWRY. The Uniqueness of Hydrogen. *Chemistry and Industry*, 1923, p. 43.

N. V. SIDGWICK. Coordination Compounds and the Bohr Atom. *J. Chem. Soc.* 1923, p. 725.

L. PAULING. *The Nature of the Chemical Bond*. New York, 1939.

LORD RUTHERFORD. The Periodic Law and its Interpretation. *J. Chem. Soc.* 1934, p. 635.

W. G. PALMER. *Valency, Classical and Modern*. Cambridge, 1944.

Annual Reports of the Chemical Society since 1904.

Chapter VI

SOME EXPERIMENTAL STUDIES ON GASES

For some three centuries gases have been the subject of experimental study at the hands of chemists and physicists, and some of the results obtained by the early investigators have assumed much importance in the light of more modern developments. Thus the simple gas laws received almost unquestioned acceptance until doubts regarding their accuracy arose in the early part of the nineteenth century. In 1827, Despretz showed that sulphur dioxide and hydrogen sulphide were more compressible than hydrogen. A number of other workers investigated the compressibility of gases, of whom particular mention should be made of Regnault in 1847, and of Amagat between 1869 and 1893. Briefly, Regnault showed that most gases are more compressible than an ideal gas which would follow Boyle's law exactly, while hydrogen was less compressible than such a gas. Amagat's experiments were conducted on a much more extensive scale, and very high pressures were employed. He was able to show that at these higher pressures all gases behaved like hydrogen, but that at somewhat lower pressures the increased compressibility noted by Regnault and others was confirmed. Thus there arose the conception of a perfect gas—a most important conception intimately connected with such subjects as the liquefaction of gases, the van der Waals equation, and the derivation of *exact* values for the molecular weights of gases by the method of limiting densities.

The development of the liquefaction of gases since the early experiments of Faraday is easy to follow. Until the year 1869 there remained a certain number of gases such as hydrogen, carbon monoxide, and nitrogen, which had resisted all attempts to liquefy them, and they were consequently designated as *permanent* gases. In 1869, however, Andrews made a very elaborate study of the isothermals of carbon dioxide, and showed that at temperatures above 31° C. no amount of pressure could effect liquefaction, while below this temperature liquefaction readily took place under an appropriate pressure. Andrews was thus able to establish the existence of a critical temperature, and the reason for designating certain gases as permanent was simply that they had never been cooled below their critical temperatures. Eventually all gases were liquefied by cooling, so that the term *permanent* as applied to gases is now of historic interest only.

Two general methods of liquefying gases have, with appropriate modifications, been employed to bring about liquefaction. There is the method of adiabatic expansion first successfully used by Cailletet in 1877 on oxygen. This method depends upon the cooling produced by the external work done by the gas during the adiabatic expansion. A different principle, namely, cooling by causing the gas to do internal work, rests upon the Joule-Kelvin effect,* discovered in 1852. Most gases other than hydrogen suffer slight cooling when allowed to escape through a fine orifice without doing external work. This cooling is a molecular effect and would not be shown by a perfect gas. In 1895, Linde, and a little later Hampson, adapted this principle in such a way as to make the cooling of the gas cumulative, and both devised apparatus for the production of liquid air in bulk.

The liquefaction of hydrogen presented exceptional difficulties for many years. Claims were made in 1883 by Wroblewski and in 1885 by Olszewski to have accomplished this, and there appears to be little doubt that they realized the production of mists or drops of hydrogen, but did not succeed in obtaining the liquid in any quantity. In 1900, Dewar showed that if hydrogen was cooled to a sufficiently low temperature, the sign of the Joule-Kelvin effect was reversed, and then the gas showed the same behaviour as other gases in passing through a fine orifice. He adapted this to the production of liquid hydrogen in bulk. In the same year a modified method of liquefying hydrogen was described by Travers. Helium proved to be still more difficult than hydrogen to be condensed to the liquid state, but the liquefaction of this gas—the last to resist—was accomplished by Onnes in 1908.

In 1898 an ingenious suggestion was made by Lord Rayleigh to improve the efficiency of the two main methods of cooling

* Often described as the Joule-Thomson effect.

gases to effect their liquefaction. He pointed out that if a gas cooled by escaping through a fine orifice were caused to work an engine, such as a turbine, a further cooling should result. The first cooling would arise as a result of the Joule-Kelvin effect, and the second in consequence of adiabatic expansion. Practical adaptations of this idea have been realized by Claude and others. In Claude's liquid air machine a considerable part of the cooling of the gas is effected by causing the issuing gas to work a reciprocating engine.

The close relationship between the gaseous and liquid states of aggregation, studied with great care on the experimental side by Andrews and a little later by Cailletet and Mathias, formed the subject of an important generalization, universally known as the van der Waals equation and put forward by its author in 1873. This equation was an attempt to apply corrections to the simple gas laws by taking account of molecular attractions and of the principles of the kinetic theory of gases. Laplace's studies on capillarity and the discovery of the Joule-Kelvin effect were the sources of the correction known as the coefficient of attraction, a. Van der Waals assumed the attraction to be inversely proportional to the square of the volume of the gas. The correction known as b is related to the diameter of the gaseous molecules. The equation thus assumed the well-known form $\left(p + \dfrac{a}{v^2}\right)(v - b) = \boldsymbol{R}\boldsymbol{T}$. Although doubts regarding the soundness of some of the assumptions which were made in formulating the equation have been entertained, everyone is agreed that the van der Waals equation is a close approximation to the truth, and it has certainly been of immense importance in various directions in chemical and physical science. Thus the conception of corresponding states has been of much value in the comparative study of the physical properties of organic compounds with a view to correlating them with the constitution. Some twenty years before, Kopp had made determinations of the specific volumes of many liquids at their boiling-points, the idea being to examine the liquids under conditions of equality of vapour pressure. The van der Waals equation has been the means of showing that such a procedure was fundamentally correct, since the boiling-points are approximately corresponding temperatures.

While the departure of gases from the requirements of the simple gas laws is the more pronounced the higher the pressures to which they are subjected, it might be expected that at extreme rarefaction Boyle's law would be closely followed; because under such conditions molecular attraction should be at a minimum, and the dimensions of the molecules would be of no account. Such is indeed the case, as was shown by some elaborate experiments carried out by Rayleigh in 1901. Between pressures of 1·5 and 0·01 mm. of mercury the product of the pressure and volume was found to be remarkably constant. This result has been of much importance in connexion with the problem of obtaining exact values of the molecular weights of gases from their densities.

The first experiments on the densities of gases to which the word *accurate* may properly be applied were those carried out by Regnault between 1845 and 1847. He introduced a counterpoise globe to eliminate the error due to buoyancy. A further and most important refinement was introduced by Rayleigh in 1888, who showed that there is an error due to loss of buoyancy when a globe is exhausted in consequence of contraction by the external pressure, and he showed how the appropriate correction could be made. Some five years later Rayleigh was able to show that the ratio of the densities of hydrogen to oxygen was 1/15·88, a number very similar to that derived from the combining ratios of the two gases. As regards conditions for comparing the densities of gases it would seem desirable that these should be made at very low pressures, because the gas laws are then closely followed. There are, however, obvious and considerable difficulties in the way of doing this *directly*. The problem of obtaining exact values of the molecular weights of gases has nevertheless been solved with great success, chiefly as the result of the work of Rayleigh, of Daniel Berthelot, and of Guye.

Between 1898 and 1907, Daniel Berthelot showed how accurate values of the molecular weights of gases could be obtained from the densities without having recourse to chemical analysis. He showed that the characteristic constants a and b of the van der Waals equation bear a definite relation to the critical constants of the gas. Corrections based upon this have enabled the densities, as determined experimentally, to be adjusted so as to correspond with the requirements of a perfect gas. Berthelot adapted the van der Waals equation so as to approximate more closely to the behaviour of gases at lower pressures. Another and more direct procedure is that known as the method of limiting densities, which has yielded excellent results in the hands of the above-named investigators. For a perfect gas Avogadro's rule should be strictly accurate, but no two gases deviate from the behaviour of a perfect gas in exactly the same way. It follows therefore that if the densities are compared under the same conditions of temperature and pressure there will be approximately, but not exactly, the same numbers of molecules in equal volumes of the two gases. The application of a correction for the coefficient of compressibility enables the approximate values to be made accurate. This is usually effected by determining the value of pv at two different pressures and then extrapolating to zero pressure.

By taking the molecular weight of oxygen as 32, Daniel Berthelot concluded that the gramme-molecular volume of a perfect gas should be 22·412 litres, and by taking account of the deviations of ordinary gases from perfection he obtained values for the gramme-molecular volume very close to this figure. Thus the values which he assigned to acetylene and to sulphur dioxide in 1904 were 22·411 and 22·417 litres respectively. Application of the method of limiting density to determine the atomic weight of nitrogen by Rayleigh, by Guye (1905), and by others led to the value of 14·009, in extremely close agreement with that obtained from the best determinations by equivalent weight methods.

In 1892, Rayleigh found that atmospheric nitrogen was invariably about one-thousandth part specifically heavier than nitrogen prepared from any of its compounds, a result which was destined to have far-reaching consequences. For some years before this, Rayleigh had given much attention to effecting various refinements in methods for the determination of gaseous densities. It appears not unlikely that he was interested in some of the experiments of contemporary workers on the combining ratios of hydrogen and oxygen, especially the volumetric experiments of Scott and of Morley; and considered that improvements in determinations of density were, perhaps, the best means of effecting progress in atomic-weight determinations of gaseous substances. The results of Scott and of Morley showed conclusively that the combining ratios of the volumes of hydrogen and oxygen were not exactly 2/1, as would have been the case with perfect gases which would have followed Avogadro's rule exactly, but very nearly (in round numbers) 2·003/1, on account of differences in the extent to which the two gases deviate from the behaviour of a perfect gas.

It was concluded that the results obtained by Rayleigh could only be accounted for by the existence of some hitherto undiscovered gas in the atmosphere, since the density determinations of nitrogen from various sources other than the atmosphere agreed well together, and the difference observed with atmospheric nitrogen was altogether outside the limits of experimental error. A discovery made by Cavendish in the year 1785, which had remained wholly neglected since that time, was recalled. Cavendish 'made an experiment to determine whether the whole of a given portion of the phlogisticated air of the atmosphere could be reduced to nitrous acid, or whether there was not a part of a different nature from the rest, which would refuse to undergo that change'. He subjected mixtures of atmospheric nitrogen with excess of oxygen to prolonged sparking over alkali to absorb the nitrous fumes, and found that a small portion of the atmospheric nitrogen resisted all attempts to cause it to combine. He remarked that 'if there is any part of the phlogisticated air of our atmosphere which differs from the rest, and cannot be reduced to nitrous acid, we may safely conclude that it is not more than 1/120th part of the whole'.

In 1894, Rayleigh found that the difference between the density of atmospheric nitrogen and nitrogen derived from its compounds was even greater than his first estimate; actually the difference was nearly $\frac{1}{2}$ per cent. It was becoming evident that the clue to the discrepancy was to be found in Cavendish's original experiment. Numerous experiments were carried out by Rayleigh in collaboration with Ramsay on attempts to isolate the unknown gas by employing various substances to absorb atmospheric nitrogen, and also by improvements on the method of Cavendish. In 1895 they announced the existence of a new gas to which they assigned the name of argon. The choice of the name was determined by the chemical inertness of the gas, but the problem regarding its nature—whether element or compound—formed the subject of an elaborate examination of its physical properties.

Values for the density of the gas were approximately 19·8, and the spectrum showed the existence of lines pointing clearly to the discovery of a new element. But the most convincing evidence regarding the nature of the new gas was derived from a study of the ratio of the two specific heats. According to the fundamental principles of the kinetic theory of gases, it had been established that the ratio of the specific heat of a gas at constant pressure to that at constant volume should be 1·67 for a monatomic gas, approximately 1·4 for a stable diatomic gas, and then gradually sink to values approaching unity for more complex molecules. These predictions received striking experimental confirmation at the hands of Kundt and Warburg in 1876, who determined the values of the ratio by measuring the velocity of sound in the gases. In particular they obtained a value of 1·67 for mercury vapour, a gas which was known on other grounds to be monatomic. Rayleigh and Ramsay accordingly determined the velocity of sound in argon and obtained a value close to 1·67, and drew the simple and obvious conclusion that the molecule of this gas is almost certainly monatomic. This conclusion was by no means readily accepted at the time. Some chemists made the far-fetched suggestion that argon might be some allotropic form of nitrogen, and certain physicists doubted the value of conclusions drawn from the elementary kinetic theory. Admittedly there were some difficulties, but the evidence in favour of regarding argon as a monatomic gas was very strong, since no gas known to be diatomic could be found with a higher ratio of the specific heats than 1·4.

The discovery of argon was soon followed by that of helium. As long ago as 1868 the existence of an element in the chromosphere of the sun was inferred by a spectral line close to the yellow lines of sodium, which did not correspond to any terrestrial element. In 1889, Hillebrand, in an investigation of certain uranium minerals, found that when uraninite was heated with sulphuric acid a considerable quantity of gas was evolved, which was not investigated very closely at the time, but was recognized as a new element by Ramsay and Travers in 1895 by its spectrum. Like argon, helium was found to be chemically inert and to have the same value for the ratio of the specific heats. A determination of the density gave figures close to a value of 4 for the atomic weight.

Having recognized the existence of two well-defined inactive elements, Ramsay and Travers considered the question as to whether they might be the only elements of such a distinctive type, or whether there might not be a group of them to be accommodated within the periodic classification. An exhaustive search for the missing gases in minerals and natural waters met with no success, but there was still the possibility that they might exist in the atmosphere. In 1898 they subjected some liquid air to fractional distillation, and obtained convincing evidence of the existence of the undiscovered gases. Between this date and 1901 they showed that there are indeed three gases in the atmosphere besides helium and argon, to which the names of neon, krypton, and xenon were assigned. This work involved experimental skill of an altogether exceptional degree, on account of the very small quantities of these gases in the atmosphere. Apart from argon which is present to the extent of about 1 per cent, the quantities of the others are measurable in parts per million of the atmosphere. The inert gases have been concerned with questions of theoretical interest which have subsequently arisen in connexion with studies such as those on radioactive phenomena, and particularly with problems of atomic structure; and some of the rare gases have assumed technical importance; neon, for example, finds application in glow lamps.

The diffusion of gases through porous septa occupied the attention of Graham for many years. His first experiments were begun in 1829, and within three or four years he had accumulated sufficient material to establish the well-known law according to which the relative rates of diffusion of gases are inversely proportional to the square roots of their densities. In 1845 he showed that the same principles apply to the *effusion* of gases through a minute orifice under a difference of pressure. Graham distinguished between what he termed the *effusion* of gases through a minute orifice of negligible length, and the *transpira-*

tion of gases through capillary tubes of considerable length. The simple square root law is not applicable to transpiration, although, speaking qualitatively, it is true that the time required for the transpiration of a light gas is less than for a heavy gas. In 1857, Bunsen adapted the effusion of gases to the determination of their densities by noting the times required for effusion of definite volumes under a definite pressure to take place; and, in 1863, Graham showed that gases of different densities could be separated by diffusion, a process which he termed *atmolysis*. The great advantage of diffusion experiments as a means of determining the density of a gas is that measurements can be made without the necessity of having the gas in pure condition. Thus in 1868, Soret determined the density of ozone by diffusion experiments. He compared the times required for ozonized oxygen, containing a known quantity of ozone, to diffuse with those required for the diffusion of chlorine similarly diluted with oxygen under conditions of strict parallelism, and his results showed clearly that the molecule of this gas is triatomic.

Diffusion and effusion methods have been of much importance in connexion with the study of the inert gases. Thus Rayleigh and Ramsay by using a train of 'churchwarden' clay pipes were able to effect a partial separation of atmospheric nitrogen into two fractions which exhibited an appreciable difference in density, and thus demonstrated the presence of a heavy gas in the nitrogen. Later, Ramsay in collaboration with Travers, applied diffusion methods for separating helium from other gases which they obtained from minerals such as cleveite. It will be recalled that Debierne in 1910 obtained a satisfactory value for the density of radium emanation by an effusion method.

In more recent times experiments on the effusion of gases have assumed increased importance. Particular interest is attached to the researches of Knudsen since 1908, who pointed out that Graham's square root law by itself furnishes no proof whatever of the correctness of the kinetic theory of gases; it could be followed if the gas were a continuum and did not consist of individual molecules. Knudsen also showed that a certain coefficient in his formula would have a value of 0·3989 if Maxwell's law of the distribution of molecular velocities is valid, whereas if all the molecules moved with the same velocity the value of that coefficient would be 0·4330. There is thus a difference of 8·5 per cent in the two values. Careful experiments to test the formula were carried out by Knudsen, and showed conclusively that the value of 0·3989 is correct, the error of experiment being of the order of 3 per cent. This furnishes a striking demonstration of a fundamental assumption of the kinetic theory of gases.

A convincing and direct experimental verification of the kinetic conception of the gaseous state was carried out by Dunoyer in 1911. He vaporized a small piece of sodium in a highly exhausted tube, divided into three compartments by partitions at right angles to the axis and perforated by small holes at their centres. The motion of the molecules of sodium was found to be chaotic in the first compartment, but a few entered the second compartment without colliding, and a fewer still reached the third compartment and made a deposit on the extremity of the tube. The selective action of the two diaphragms was therefore to cause a very small proportion of the molecules to move like a ray. The deposit obtained by these molecular rays of sodium, as they have been termed, was exactly of the type to be expected from the geometrical dimensions of the apparatus.

Since the pioneer work of Dunoyer, a great deal of experimental work on molecular rays has been carried out, particularly since 1926 by Stern, Gerlach and others. Much of the success of work of this kind has resulted from developments of modern methods of obtaining extremely high vacua. Thus by using a method which was an imitation of Fizeau's toothed wheel experiment for measuring the velocity of light, Lammert in 1929 succeeded in verifying Maxwell's law for mercury atoms.

While the passage of gases through porous materials is in general determined solely by the density of the gas and not by other properties, a few examples are known of materials which allow some gases to pass through them but resist the passage of others. As long ago as the year 1829, Graham observed that when an india-rubber bladder was partly filled with coal gas and then completely immersed in an atmosphere of carbon dioxide, the bladder became fully inflated. It followed that india-rubber is permeable to carbon dioxide but not to other gases, and even at that early date Graham ventured a comparison with a similar phenomenon which had been observed long before by Dutrochet with dissolved substances and termed *endosmose*. In 1866, Graham observed the remarkable phenomenon of the occlusion of hydrogen by palladium, and devoted the next three years to a thorough study of *hydrogenium*, as he termed the gas thus alloyed in the metal. The nature of the product obtained when hydrogen is absorbed by palladium has been the subject of numerous later investigations; but for the present purpose attention may be directed to an interesting experiment which was carried out by Ramsay in 1894, which has a very direct bearing upon the problems connected with the selective capacity of materials for permitting the passage of certain gases through them. Ramsay's experiment was an attempt to obtain some convincing demonstration of the analogy, regarded at that time by many as much more than a merely superficial one, between gaseous pressure and osmotic pressure. Briefly the experiment consisted in having a vessel divided into two equal compartments by a wall of palladium, one compartment being filled with nitrogen and the other with hydrogen, both being at the same pressure. On raising the temperature of the apparatus the palladium permitted the passage of the hydrogen but not of the nitrogen, and on cooling to the original temperature the pressure in one compartment was reduced to one-half and in the other compartment it was increased to $1\frac{1}{2}$ times its original value. The excess of pressure in one compartment might thus be regarded as the 'osmotic' pressure of the nitrogen. Another example of selective permeability to certain gases was observed by Jaquerod and Perrot in 1905. They found that quartz at high temperatures permitted the passage of helium and of hydrogen but not of other gases. Subsequent experimentalists have made some use of diffusion through quartz as a means of separating helium and neon, but other methods of effecting this object have been found more satisfactory.

Many reactions between gases are profoundly influenced by changes in the physical conditions, particularly as regards the presence or absence of catalysts, and in no field of investigation is this of greater interest or importance than in the study of combustion. This subject can be dated as far back as the early years of the nineteenth century associated particularly with the names of Davy and Faraday. Davy was familiar with the catalytic properties of platinum in inducing the combination of hydrogen and oxygen, and he realized that these gases could be caused to combine slowly by contact with the metal at lower temperatures than that of the very ill-defined ignition point. The flameless combustion of alcohol vapour in contact with a spiral of platinum wire was observed in 1819 by Döbereiner, who invented a lamp based upon this principle. Shortly after that time Faraday and de la Rive put forward different views regarding the nature of the catalytic process. Faraday regarded the phenomenon as due to the occlusion of the gases by the metal, the idea being that under such conditions the gases were in a condition comparable with being under very high pressure, and therefore presumably more reactive. De la Rive, on the other hand, considered the process to consist in a rapidly alternating series of oxidations and reductions at the surface of the metal. In the century which has passed since Faraday's time it will be seen that both theories of the catalytic process have found favour with different investigators, and even at the present time there is no general agreement, but the alternative views are each entitled to consideration.

After Faraday's work, the subject of gaseous combustion received much study at the hands of Bunsen, and the invention of the familiar burner associated with his name was made in 1855. Shortly after this Bunsen made some experiments on the speed with which flames are propagated in gaseous mixtures—a subject which has subsequently been studied in very great detail. Of somewhat greater value were some experiments which Bunsen carried out about the year 1867 on the pressures produced in explosions. The further development of the study of flames and of gaseous explosions was carried out chiefly by Berthelot and Vieille, by Mallard and Le Chatelier, and especially by Dixon. The subject was approached from the standpoint of thermochemistry by Berthelot, who conducted a lengthy series of researches chiefly between 1865 and 1880. One of the very valuable practical results of this work was the invention of his bomb calorimeter, which has been of the utmost value for the determination of heats of combustion. On the theoretical side Berthelot's thermochemical studies were less successful, as it was during this time that he put forward his celebrated law of maximal work. According to this principle Berthelot asserted that every chemical change which takes place without the application of external energy gives rise to the production of those substances the formation of which is attended with the maximum evolution of heat. Although the statement is very frequently true, it has been shown to be unsound both from the thermodynamic and from the experimental standpoint. To Berthelot and Vieille and to Mallard and Le Chatelier belongs the credit of having followed up the development of the subject of explosions by Bunsen, especially the discovery of the explosive wave.

Although an ignition point was, in an ill-defined way, recognized by the earlier investigators, its determination for various gaseous mixtures was no simple matter. There is no difficulty in heating up a gaseous mixture and noting the temperature at which inflammation or explosion takes place, but such a temperature does not admit of precise characterization, because it must depend upon the amount of chemical combination which has taken place before this supposed ignition point was reached. An important step towards getting this temperature determined with greater exactness was taken by Dixon and Coward in 1909, who heated the combustible gas and air or oxygen separately before allowing them to mix. Consistent results seem to have been attained in this way, but it must nevertheless be still somewhat uncertain as to whether the term *ignition point* is really capable of strict definition as a physico-chemical property. Another interesting method of attacking this problem was devised in 1907 by Falk as a result of a suggestion due to Nernst. Falk's method consisted in igniting the gases by adiabatic compression, which was effected by confining them in steel cylinders and applying sudden pressure by allowing a weight to fall on a piston. Although subsequent experiments carried out by Dixon between 1910 and 1914 seem to confirm Falk's contention that the heating of the gaseous mixtures was effected solely by adiabatic compression until the ignition point was reached, some of his other assumptions seem to have been altogether unjustified. The ignition of gases by adiabatic compression has, however, attracted attention from Tizard and Pye and from others since

1922 in connexion with problems of internal combustion engines.

Among the numerous problems connected with slow combustion, one of considerable interest since the time of Bunsen was finally solved by the investigations of Bone and his collaborators between 1898 and 1912. This was the question whether in the combustion of a hydrocarbon there is any preferential combustion of either carbon or hydrogen. Many chemists favoured such a view, and it may be said that on the whole the idea of preferential combustion of hydrogen was the more popular, but advocates of the doctrine of preferential combustion of carbon were not wanting. Other chemists considered that the process of oxidation involved simultaneous attack upon both constituents. Bone first tried the effects of heating the gaseous mixtures in sealed glass bulbs to temperatures below the ignition point, and testing the products of the reaction by analysis. The method of heating in closed bulbs had previously been employed by Victor Meyer to determine ignition points, but with necessarily imperfect results. These experiments with closed bulbs as carried out by Bone demonstrated the very important fact of the formation of aldehydes during the slow oxidation. The production of aldehydes as a necessary part of the mechanism, as distinct from being by-products, was shown by the use of what Bone termed his circulation apparatus. This apparatus consisted of a closed system for causing the gaseous mixtures to be circulated over heated catalysts by the action of automatic mercury pumps, and provided with an arrangement for removing any condensable products for analysis. With this apparatus Bone was able to show that the attack on the hydrocarbon molecule by oxygen almost certainly takes place by a process of hydroxylation, followed by the decomposition of the product into an aldehyde and steam. Any theory of preferential combustion of carbon or hydrogen was ruled out; the oxidation must involve simultaneous reaction with both elements.

The numerous studies on the effect of moisture on combustion may be said to have begun in 1880 with Dixon's experiments on the reaction between carbon monoxide and oxygen. This reaction was found to be seriously impeded by the absence of water vapour, and a fair degree of drying was found to inhibit oxidation altogether. This subject was then studied with great care by Baker who mastered the difficult technique of drying gases with phosphorus pentoxide to a very high degree, and most gaseous mixtures prepared in this way were found to be remarkably inert to electric sparks. Actually the carbon monoxide-oxygen mixtures were found to require a considerably lower degree of desiccation than others so as to become resistant to the action of the spark. As regards hydrocarbons, most of Bone's experiments seemed to show that the presence or absence of moisture made little if any difference to their sensitiveness to undergo oxidation. This has raised the difficult and controversial question as to whether pure substances are chemically non-reactive, but require some 'impurity' to act as a catalyst so that reaction may take place. As regards carbon monoxide the question of the necessity of the presence of water vapour to bring about inflammation must certainly be answered in the negative as the result of some very interesting experiments carried out by Bone and his collaborators between 1923 and 1925.

Bone had noted that certain investigators, e.g. Thornton (1914–16), in studies on the ignition of inflammable gaseous mixtures had found that what might be termed a certain 'minimum spark energy' was required to bring about explosion, and considered that some of the results observed by Dixon with well-dried mixtures of carbon monoxide and oxygen might be due to the use of insufficiently powerful sparks. In other words, Bone decided to carry out experiments on highly dried mixtures using sparks of gradually increasing energy to see whether ignition would eventually take place. The experiments were begun by determining the minimum condenser discharge necessary to ignite mixtures of carbon monoxide and oxygen (two volumes of the former to one of the latter), beginning with mixtures (a) saturated with water vapour, (b) dried with calcium chloride, and (c) intensely dried with phosphorus pentoxide. The results obtained by plotting the percentage of water vapour by volume in the gaseous mixtures against the condenser capacity required for ignition were found to fall on a smooth curve. Eventually Bone found that these gaseous mixtures when dried for periods of the order of six months by phosphorus pentoxide could be exploded by condensers of the capacity of about 1 microfarad charged at 970 volts. The idea of the necessity for the presence of water vapour as a catalyst to bring about explosion had perforce to be abandoned, and in view of modern work on the kinetics of reactions the importance formerly attached to this subject would now appear to be altogether exaggerated. Nevertheless, the acceleration of the reaction between carbon monoxide and oxygen by water vapour is particularly interesting as the mechanism seems to be different from that of the direct oxidation of the dry gas. Dixon in 1884 concluded by regarding the oxidation of moist carbon monoxide to take place in a manner represented by the consecutive reactions: $CO + H_2O = CO_2 + H_2$ and $2H_2 + O_2 = 2H_2O$. Modern studies of the kinetics of this reaction have shown that it proceeds by a chain mechanism.

It should be noted that many reactions between gases which were formerly considered to be homogeneous are in reality not so, and too much attention has been paid to the degree of desiccation of the gases and far too little to the nature of the walls of the apparatus within which the gases react. The study of the kinetics of gaseous reactions has made this abundantly clear. One of the first things to be done in studying any gaseous reaction from the standpoint of kinetics is to determine whether the reaction is a truly homogeneous one or a 'wall' reaction. One single example of this may be quoted. In an interesting study of the reaction between ethylene and bromine, Norrish in 1923 showed that the rate with which these gases unite is profoundly modified by the nature of the surface of the reaction vessels. He found that the gases react readily within surfaces of glass, whether plain or coated with stearic acid or with cetyl alcohol, but when coated with paraffin wax the reaction was brought almost to a standstill.

Of the numerous gaseous reactions which are catalyzed by metals or other solids, a brief reference must be made to one or two examples of outstanding importance. The union of sulphur dioxide with oxygen in the presence of platinum to form sulphur trioxide was known to Davy, and in 1831 Peregrine Phillips, a vinegar manufacturer in Bristol, actually started a plant for making sulphuric acid for his requirements by what is essentially the contact process. Although he soon found that the platinum gradually lost its efficiency as a catalyst, Phillips should be considered as the true inventor of the contact process: indeed, he found that to obtain the best results a considerable excess of atmospheric oxygen was necessary, and this was some thirty years before the recognition of the law of mass action. The subsequent history of this most important manufacturing process has been very largely concerned with the necessity of obtaining the gases in a purified condition before entering the contact chamber, because certain impurities, particularly arsenic, cause very great diminution in the catalytic activity of the platinum.

Numerous reactions between organic compounds in the gaseous condition have been effected with the aid of catalysts, particularly nickel. This subject has received a great deal of study at the hands of Sabatier and his collaborators, especially Senderens and Mailhe, between 1899 and 1919, who favoured a theory based upon the formation of intermediate compounds. Thus it was found that a mixture of carbon monoxide and hydrogen can be quantitatively transformed into methane and steam by passing the gases over nickel at about 200° C. Numerous dehydrogenations of organic compounds have been effected by passing the vapours over heated metals. For example, primary alcohols were converted into aldehydes and hydrogen by Sabatier and Senderens (1905) by passing the vapour over heated copper at about 250° C., and secondary alcohols under similar conditions are decomposed into ketones. The catalytic action of certain metallic oxides was found to be very different from that of metals. Thus it was found that whereas metals bring about the dehydrogenation of alcohols, certain metallic oxides, particularly anhydrous alumina, cause the withdrawal of the elements of water. Thus when the vapour of ethyl alcohol is passed over heated aluminium oxide either ethylene or ether can be formed according to the conditions. Sabatier and Mailhe in 1910 found that thorium oxide and the blue oxide of tungsten were also very efficient for preparing olefines from alcohols.

The absorption of gases by liquids was studied by some of the earlier investigators, particularly by Cavendish, Henry, and Dalton. In 1803, Henry showed that the mass of a sparingly soluble gas which is absorbed by a liquid is proportional to the pressure. This law was then extended by Dalton to mixtures of gases, who showed that each gas is absorbed proportionally to its partial pressure. Later, Bunsen devised a special absorptiometer for determining the solubility coefficients of sparingly soluble gases, with which he verified the simple laws more exactly. This apparatus, together with numerous exact methods of gas analysis, was described in the *Gasometrische Methoden* in 1857. The solubility coefficients were determined by placing measured volumes of the gas over mercury and adding a volume of the liquid. The open end of the tube was then screwed down against an india-rubber plate, and the whole apparatus shaken violently to saturate the liquid with the gas. After saturation had been attained, the solubility was measured by the contraction in volume of the gas which remained undissolved. A somewhat more convenient form of absorptiometer was later introduced by Ostwald, who expressed his results by a simpler absorption coefficient, namely, by the ratio v/V, where v is the volume of the gas which is absorbed by a volume V of the liquid.

The simple laws discovered by Henry and Dalton are by no means followed by the very soluble gases, where some form of chemical combination must be considered to take place. Thus anhydrous hydrogen chloride, a compound devoid of acid properties, becomes an extremely strong acid when dissolved in water. In other cases evidence of chemical combination is less immediately obvious. Thus with ammonia, although the greater part of the gas appears to dissolve 'unchanged', there is definite evidence of the presence of hydroxyl ions in solution, so a certain small proportion is presumably present as ammonium hydroxide. Somewhat similar considerations are applicable to solutions of carbon dioxide, the feebly acidic character of which affords evidence of the existence of the unstable carbonic acid in solution.

Throughout a large part of the nineteenth century it was tacitly assumed that constancy of composition afforded definite evidence of chemical combination. As most gases can be completely removed from solution by boiling, their solutions were consequently regarded as physical mixtures. The behaviour of the solutions of the halogen hydrides is, however, quite different. Before the year 1860 it was considered that a solution of hydrochloric acid containing 20·2 per cent of hydrogen chloride was a definite compound, because an acid of this composition was obtained whenever hydrochloric acid of *any* concentration was distilled under atmospheric pressure. In that year, however, it was shown by Roscoe and Dittmar that if the external pressure was varied the composition of the acid which distilled when a steady state was reached was different. In this way the existence of constant boiling mixtures became recognized, and it was made perfectly clear that in order to characterize a material having a definite composition as a compound it must preserve this constancy of composition over a certain range of conditions. This whole subject was expressed in a more generalized form by Willard Gibbs between 1875 and 1878, who stated a theoretical principle, known as the phase rule, which deals with equilibrium in heterogeneous systems. Willard Gibbs derived the phase rule from thermodynamic considerations, but the elaborate mathematical methods which he used may perhaps account for its neglect for some ten years until Roozeboom pointed out its value in classifying chemical equilibria. Thus the principles of the phase rule have been found of much value in the study of equilibria between gases and solids.

It has been well established by numerous experiments that when a gas is taken up by a solid there are three different ways in which this may happen. The gas may dissolve in the solid in accordance with Henry's law, the amount dissolved being directly proportional to the pressure. Or the gas may be absorbed by the solid, and the pressure remain constant over a certain range of concentration of the gas in the solid. In this way the existence of what in the language of the phase rule is termed a univariant system, consisting of two solid phases and a gas, becomes apparent, indicating the formation of a compound. The equilibrium between calcium carbonate, lime, and carbon dioxide was first studied by Debray as long ago as 1867, who observed that the pressure of the carbon dioxide was a function of the temperature only, and was independent of the amounts of lime and of calcium carbonate. More accurate measurements of this dissociation pressure were obtained by

Le Chatelier in 1883. Many other experimental studies of solid-gaseous systems have since been made, and the principles of the phase rule have been regularly applied to ascertain whether the pressure-concentration relations indicate the formation or non-formation of compounds. A third way in which a gas may be taken up by a solid is for the concentration of the gas in the solid phase to vary according to some power of the pressure. This is known as the adsorption-product type, and is well illustrated by the phenomena observed when gases are occluded by charcoal.

Although the three ways by which a gas may be taken up according to the principles of the phase rule have been well defined, experimental work on this subject has frequently been found somewhat difficult. Thus it has sometimes happened that equilibrium has been difficult to establish on account of the slowness with which gases diffuse into the interior of solids. Very numerous experiments have been made on the occlusion of hydrogen by palladium. In 1895, Hoitsema acting on suggestions due to Roozeboom determined the pressure-concentration relationships of this system. The results were far from simple to interpret, but the general conclusion was reached that although some indications of a univariant system were obtained, it was more likely to be due to the formation of two immiscible solid solutions than to the formation of a definite compound. The hydrides of the alkali metals were studied by Troost and Hautefeuille in 1874, who considered them to be represented by the formulae Na_2H and K_2H. In 1902, Moissan formulated them as NaH and KH respectively, and regarded the results observed by Troost and Hautefeuille as due to the formation of solid solutions of the metals and the hydrides. In 1921 the hydrides of the alkali and alkaline earth metals and also of certain of the rare earths were investigated more fully by Ephraim and Michel, and it is clear that the compounds had previously been obtained in scarcely pure condition. They were able to obtain fairly reliable values for the dissociation pressures, and incidentally to show that the formulae of the alkali, alkaline earth, and rare earth hydrides were of the RH, RH_2 and RH_3 type respectively.

The analysis of gaseous mixtures by methods involving

measurements of changes of volume may be said to have begun with Cavendish. It was later, between 1838 and 1880 and especially about 1856, brought to a high degree of accuracy by Bunsen. Two general methods of effecting the necessary chemical changes were used, namely, explosion and absorption with reagents. Regarding absorption methods, Bunsen's procedure consisted of applying the solid reagents in the form of pellets attached to platinum wires to the gaseous mixtures, and observing the resulting diminution of volume. Although the results attainable by this method were remarkably accurate the method was extremely slow. The introduction of liquid reagents by Williamson and Russell in 1864 was a distinct advance; but it is interesting to note that Bunsen realized the possibility of using reagents in solution, but decided against so doing because of the errors necessarily introduced on account of the solubility of the gases in the liquid solvents. When as in most work rapidity is scarcely less important than extreme accuracy, the advantages of liquid reagents are very obvious, and their employment is now almost universal in all ordinary work where moderate volumes of gas are being analysed, as in the well-known methods devised by Hempel in the latter years of the nineteenth century.

Coming to more recent years, the outstanding advances on the experimental side have been the development of methods and of apparatus for dealing with extremely small quantities of material—an advance which may truly be said to apply to all practical work in chemistry. The invention of Sprengel's mercury pump in 1865, followed at a later date by the more convenient pump devised by Töpler, enabled experimentalists to produce very high vacua, and investigate the behaviour of gases at very low pressures. The Töpler pump was constantly used by Ramsay and his collaborators in work on the inert gases, and one of the most striking experimental triumphs realized by Ramsay and Soddy was their determination of the volume of radium emanation derived from radium bromide in 1904, in which the volume of gas measured was a very small fraction of a cubic millimetre. They estimated the volume of emanation in equilibrium with 1 g. of radium to be about 1 cu. mm. In 1908, Rutherford obtained a more accurate value, viz. 0·6 cu. mm., which has been verified by others. The analysis of the commoner gases by explosion and absorption methods has passed from the macro to the micro scale, changes of volumes being measured with burettes having capillary tubes, as, for example, in the apparatus devised by Christiansen in 1925. In these micro-methods for gas analysis the use of solid absorbents attached to platinum wires as originally due to Bunsen has returned, and a method for absorbing gases devised by Dewar in 1905 using charcoal cooled in liquid air has been used with much success in fractionating gases, and thus determining their proportions.

The production of high vacua by mercury pumps and by other methods has provided the means of studying the behaviour of gases at extremely low pressures, and some of the results obtained in this way have been of much importance in very various directions. In 1873, Crookes was engaged on an elaborate research on the atomic weight of thallium, and in order to obviate the necessary calculations for reducing the weighings to vacuum standard he used a balance inside a highly exhausted case. In carrying out the weighings some curious effects were observed which were subsequently traced to the repulsive action of radiant heat, and which ultimately led to the invention of the well-known radiometer. A few years later Crookes began the celebrated experiments on the cathode-rays which involved work at the lowest pressures then attainable, and in 1881 he succeeded in showing that the viscosity of gases, which for wide ranges of pressure is independent of the pressure, falls off very rapidly as the pressure is reduced to extremely low values. It had been pointed out by Maxwell that according to the kinetic theory of gases, physical properties such as viscosity and thermal conductivity should be independent of the pressure over very considerable ranges of pressure, but such a state of affairs could no longer continue if the pressure were reduced to such an extent that the mean free paths of the molecules were comparable with the distances between the walls of the apparatus. Crookes's observations on viscosity have been amply verified by later investigators. As regards thermal conductivity, it was shown by Soddy and Berry in 1910 that at extremely low pressures this property is also proportional to the pressure. Much work, both theoretical and experimental, has since been done on this subject, especially by Knudsen, who introduced an important conception, known as the coefficient of accommodation, to denote the incompleteness of interchange of energy between the walls and the gas molecules.

Of greater interest to chemists are some experiments carried out by Langmuir since 1912 on the effect of electrically heated tungsten, platinum, or palladium wires in gases at pressures of 0·01 mm. or less. He found that if tungsten filaments are heated to high temperatures in hydrogen at these low pressures, the gas is partly dissociated to the atomic condition, the process appearing to consist in the gas dissolving in the metal in the atomic state and then diffusing out again, without much recombination to form molecules taking place. This subject was further studied by Wood in 1921–2 and by Bonhoeffer in 1924, who devised an improved method of preparing atomic hydrogen. The gas is very different from ordinary molecular hydrogen, being very much more active chemically. Thus it combines with many elements directly at the ordinary temperature forming hydrides. Several experimenters have obtained some evidence of the formation of a triatomic hydrogen by subjecting ordinary hydrogen to electric discharges at low pressures, and have named the product *hyzone*, as ozone is formed from oxygen under somewhat similar conditions. An active modification of nitrogen was prepared by Strutt in 1911 by subjecting the gas under low pressures to a jar discharge, and although much experimental work has been done on this product it appears uncertain whether this gas is atomic nitrogen or has some different constitution. It is in any case highly reactive. All attempts to obtain atomic nitrogen by thermal methods have so far been unsuccessful.

It should be emphasized that the success realized by Langmuir and others in preparing atomic hydrogen was due to the removal of the atoms to a distance so as to avoid recombination with other atoms. This principle, namely, the removal of the products resulting from thermal dissociation so as to avoid recombination as far as possible, has been used at various times ever since the year 1847, when Grove showed that water vapour could be partially dissociated by an intensely heated platinum wire or by the agency of electric sparks. About 1864, Deville devised his celebrated apparatus, usually known as the hot-cold tube, for the study of gaseous dissociation. The gases were passed through a white-hot porcelain tube having an inner tube of silver through which a stream of cold water was allowed to flow. In passing through the annular space between the two tubes, the gas was caused to dissociate by contact with the intensely heated outer tube, but by diffusing towards the cold wall of the inner tube the sudden cooling impeded the reformation of the original compound to a considerable extent. In this way Deville was able to demonstrate the thermal dissociation of gases such as carbon dioxide, sulphur dioxide, and hydrogen chloride. For more exact studies on the degree of dissociation and of chemical equilibria in gaseous systems in general, Nernst made various improvements in Deville's original method, and in conjunction with von Wartenberg in 1906 obtained values for the degree of dissociation of steam at various high temperatures in good agreement with the values calculated from the equation connecting the equilibrium constant with the absolute temperature, usually known as the reaction isochore $\dfrac{d \log K}{dT} = \dfrac{Q}{RT^2}$. About the same time and in conjunction with Jellinek, Nernst studied the equilibrium between nitric oxide and its elements at various high temperatures and the velocity of the reaction under those conditions. The values obtained for the percentage yield of nitric oxide at various temperatures were in good agreement with what should be expected theoretically.

The preparation of atomic hydrogen was followed in 1929 by that of the free unstable radicals methyl and ethyl by Paneth and his collaborators, and here again success was due to rapid removal of the products of thermal dissociation to a distance. Throughout the last half of the nineteenth century, and indeed for some time afterwards, chemists were almost unanimously agreed that organic radicals such as methyl and ethyl could have no free independent existence: the question as to whether there might not be the possibility of their having a very *transitory* existence was regarded as either highly improbable, or in any case too difficult to be investigated experimentally. In 1929, however, it was shown by Paneth and Hofeditz that when the vapour of lead tetramethyl was heated in a stream of a gas such as hydrogen for carrying the dissociation products to a distance, the production of free methyl radicals was clearly demonstrated, as was shown by their action upon metallic mirrors deposited on the glass. In 1931, Paneth and Lautsch prepared the free ethyl radical in a similar way by the thermal dissociation of lead tetraethyl using hydrogen as the carrying gas. Paneth and his collaborators were able to determine the half-life of these unstable alkyl radicals under specified conditions, by observing the times required for them to remove metallic mirrors, placed at definite distances from the source of preparation, when carried by a gas at a known rate. In this way they showed that the half-life is extremely short—of the order of a few thousandths of a second. It is interesting to note that methyl and ethyl are the only free radicals which have been obtained in this way. Paneth and Lautsch in 1931 showed that radicals higher than ethyl are extremely unstable, since it was found that methyl radicals were the chief products found in the thermal decomposition of lead tetra-n-propyl and lead tetra-iso-butyl.

The study of free radicals has resulted in obtaining a clearer understanding of some reactions which have long been known. Thus the well-known method for preparing hydrocarbons such as ethane by the action of sodium upon methyl iodide, due to Wurtz, has assumed a renewed interest as a result of experiments carried out by Polanyi and others since 1930. These experiments consisted in carrying sodium vapour, at a pressure of the order of magnitude of one-thousandth of a millimetre, by a current of a non-reactive gas into a reaction chamber containing the alkyl bromide at a pressure of a few hundredths of a millimetre. Under these conditions clear evidence was forthcoming that the reaction $Na + CH_3Br = NaBr + CH_3$ took place, and measurements of the speed of the reaction were carried out by following the rate of disappearance of the sodium vapour. The formation of the free methyl and ethyl radicles was proved by introducing iodine into the apparatus, and subsequently identifying the methyl or ethyl iodide thus formed. As the free radicals obtained in this way were carried to some distance from their source, it is clear that an alternative method of preparation to that of Paneth has been made available.

It is instructive to compare the modern views regarding free radicals with those which prevailed during the nineteenth century. Chemistry as understood by such men as Frankland, Couper, or Kekulé, afforded no possibilities for the existence of free radicals. Much more enlightened, however, were the views of Nef. In elaborating his views on bivalent carbon, Nef on more than one occasion between 1897 and 1904 considered that the methylene radical must have a transitory, but none the less definite, existence. He thought that many of the reactions of such compounds as methyl alcohol or methyl chloride could best be explained by the molecules of these compounds decomposing in such a way as to produce a very small proportion of free methylene. Although Nef did not succeed in establishing his ideas on an experimental basis, it is very interesting to note that Rice and Glasebrook in 1933 were able to demonstrate the formation of free methylene in the thermal decomposition of diazo-methane. It is thus clear that if the older generation of chemists had but refrained from saying that free radicals *could not* exist, but had been content with some more cautious or reserved form of expression, such as 'free radicals are groups of atoms which have no *stable* existence as such', their views would have been much closer to those which are held at the present time.

REFERENCES

T. GRAHAM. *Chemical and Physical Researches*. Edinburgh, 1876.

R. W. BUNSEN. *Gasometrische Methoden*. English translation by H. E. Roscoe. London, 1857.

M. W. TRAVERS. *The Experimental Study of Gases*. London, 1901.

M. W. TRAVERS. *The Discovery of the Inert Gases*. London, 1928.

P. SABATIER. *Catalysis in Organic Chemistry*. Translated by E. Emmet Reid. London, 1923.

A. FARKAS and H. W. MELVILLE. *Experimental Methods in Gas Reactions*. London, 1939.

W. A. BONE. Fifty Years Research upon the influence of Steam on the Combustion of Carbonic Oxide. Third Liversidge Lecture. *J. Chem. Soc.* 1931, p. 338.

J. H. JEANS. Van der Waals Memorial Lecture. *J. Chem. Soc.* 1923, p. 3398.

SIR JAMES JEANS. *An Introduction to the Kinetic Theory of Gases*. Cambridge, 1940.

M. KNUDSEN. *The Kinetic Theory of Gases*. London, 1934.

R. G. J. FRASER. *Molecular Rays*. Cambridge, 1931.

F. O. and K. K. RICE. *The Aliphatic Free Radicals*. Baltimore, 1935.

Annual Reports of the Chemical Society since 1904.

Chapter VII

SOME PROBLEMS OF SOLUTIONS

The question regarding the nature of a dissolved substance—whether a mere physical mixture or in some manner chemically combined with the solvent—has occupied much of the attention of chemists at different times. The modern study of this subject may be said to date from about the year 1885, when van't Hoff published his celebrated memoir entitled *Lois de l'Équilibre dans l'État dilué, gaseux ou dissous*, the title of which was obviously intended to emphasize the condition of a dissolved substance as being similar to that of a gas. The foundations of van't Hoff's theory of dilute solutions were the osmotic experiments of certain plant physiologists, especially the quantitative measurements of Pfeffer carried out in 1877, and the experiments with plant cells on isotonic solutions of de Vries in 1884, and also the experiments of Raoult on the freezing-points and vapour pressures of dilute solutions about the year 1882. Van't Hoff was able to show that the gas laws were applicable to dilute solutions, and in particular that Avogadro's rule was valid in the sense that one gramme-molecular weight of a solute (non-electrolyte) dissolved in a total volume of 22·4 litres produces an osmotic pressure of one atmosphere at 0° C. If that weight of this substance could have existed as a gas under those conditions of temperature and volume the value of the gaseous pressure produced would have been the same as that of the osmotic pressure, and on this equality van't Hoff drew the simple but unjustified conclusion regarding the identity of the nature of gaseous and osmotic pressures.

In discussing this difficult and exceedingly controversial question it should never be forgotten that although accurate methods of measuring osmotic pressure are available, all such methods depend ultimately upon the balancing of the osmotic pressure by a hydrostatic pressure, and it is the latter which is measured. No information whatever regarding the nature of the osmotic process or the mode of action of the semi-permeable membrane is obtainable from such measurements, any more than is information forthcoming regarding the nature of gravitational attraction from measurements of the weights of objects made with a spring balance. In saying this, however, there is not the slightest intention of minimizing the very great practical value of the results which have arisen in consequence of experimental studies on osmotic phenomena; particularly the valuable methods of determining molecular weights in solution by the measurement of properties such as the depression of the freezing-points or of the elevation of the boiling-points of solutions. These practical methods are really derived from van't Hoff's theoretical proof that the lowering of the vapour pressure experienced by a solvent on adding a non-volatile solute is equal to the ratio of the number of molecules of the solute to the number of molecules of the solvent, or as it is usually expressed symbolically,

$$\frac{\Delta p}{p} = \frac{n}{N}.$$

Although the theory of dilute solutions was derived from the application of the principles of thermodynamics to the study of osmotic phenomena, and the nature of the experimental results obtained seemed to suggest explanations based upon some theory of direct molecular bombardment by the solute, in a manner analogous to the phenomena of gaseous pressure being due to bombardment by the molecules of the gas an alternative explanation is possible. It was shown by Poynting in 1896 that the equation $\frac{\Delta p}{p} = \frac{n}{N}$ could be derived from a theory based upon some form of chemical combination between the solute and the solvent. Very briefly Poynting's explanation consisted in assuming that the molecules of the solute form non-volatile compounds with a certain proportion of the molecules of the solvent. This may be expressed by saying that each molecule of the solute diminishes the facility for evaporation of x molecules of the solvent by $1/x$th part. Denoting as before the numbers of solute and solvent molecules by n and N, it will be evident that there will be $N - n$ solvent molecules left unaffected and therefore available for evaporation, but N solvent molecules will be available for condensation. The vapour pressure will therefore be reduced in the ratio of $(N-n)/N$, from which it follows that $\dfrac{\Delta p}{p} = \dfrac{n}{N}$. It will be evident that a theory of this kind can be readily extended to include cases in which the molecules of the solute are dissociated into ions.

The problem of the condition of dissolved substances has been approached from other angles. Thus, in 1872, Berthelot and Jungfleisch studied the distribution of a solute between two immiscible solvents. As the result of a number of experiments it was concluded that generally speaking a solute distributes itself between the solvents in a constant ratio, usually termed a partition coefficient, which is the ratio of the solubilities of the substance in the two solvents. This subject was also studied by Nernst in 1891, who showed that some of the irregularities which had been observed by Berthelot and Jungfleisch were due to differences in the molecular condition of the solute in the two solvents. A simple partition coefficient is only obtained when there is identity in the molecular condition. In other cases more complex relations were found to exist, as for instance in the distribution of benzoic acid between water and benzene, double molecules being formed in the latter solvent. Another and very important avenue of approach is by way of the study of the colour and absorption spectra of solutions. A great deal of work has been done on this subject, and although some of the results have been capable of interpretation in more than one way, it has been possible to supplement them by other methods of inquiry. An instructive example is to be found in the case of iodine. This element dissolves in a number of organic compounds and in aqueous solutions of alkali iodides. In general, two differently coloured types of solutions are formed. In certain types of organic solvents, such as hydrocarbons, chloroform, and carbon disulphide, the colour is violet, whereas in compounds such as alcohol, ether, or pyridine the solutions are coloured brown. In the aqueous solutions of inorganic iodides the colour is also brown. The molecular weight of the element as determined in these solutions by the cryoscopic method, when applicable, seemed to show that there were no great differences in the molecular weight which usually corresponded to a diatomic molecule. Jakowkin between 1894 and

1896 studied the partition coefficient of iodine between immiscible organic solvents and water and solutions of potassium iodide. From his results he concluded that the brown colour of the latter is due to the presence of a complex anion I_3^-. Further work by other investigators was summarized by Waentig in 1909 who concluded, largely as a result of studies on the absorption spectra, that in the violet solutions the element is present in an 'unchanged' condition as the diatomic molecule, while in the brown solutions there was a certain amount of combination between the solute and the solvent. This is consistent with the ease with which iodine can be extracted from its solution in an iodide by solvents such as chloroform.

Although van't Hoff's principal contribution to the theory of dilute solutions was his establishment of the analogy between osmotic pressure and gaseous pressure, he nevertheless recognized that velocity of reaction and equilibrium may be affected by solvents. In 1898 he showed theoretically that the ratio of the concentrations of two dynamic isomerides of a substance present at equilibrium in any particular solvent should be in the ratio of the solubilities of these isomerides in that solvent, or if A and B are the two isomerides the result may be expressed thus:

$$\frac{\text{Concentration of } A}{\text{Concentration of } B} = \frac{\text{Solubility of } A}{\text{Solubility of } B} \times G,$$

where G is a constant which depends solely upon the nature of the two compounds and upon the temperature. This important principle was verified experimentally by Dimroth between 1904 and 1913 for certain desmotropic substances such as the dynamic isomerides:

A hydroxy triazole carboxylic ester   A diazo malonic ester amide

Dimroth was able to obtain some further experimental evidence in support of this general principle, and he emphasized what he regarded as an important distinction between the physical and the chemical properties of solvents. To the former he assigned those properties concerned with the molecular weights of the dissolved substances, in short with colligative properties in which the specific properties of the solvent are not concerned. The chemical properties were to be regarded as definitely specific and closely connected with its action as a solvent.

Of the various methods for determining molecular weights in solution which have arisen as a result of the work of Raoult and van't Hoff, the cryoscopic and ebullioscopic methods have been found of the greatest practical value, particularly the former. Since the year 1889 Heycock and Neville extended cryoscopic methods to determine the molecular weights of a number of metals when dissolved in a second (solvent) metal. Thus by using tin as the solvent, they determined the freezing-points of alloys with metals such as silver, gold, copper, magnesium and others. In many cases the values which were obtained for the molecular weight of the solute were identical with that of the atomic weights of the metals, but in the cases of aluminium and indium there were indications of more complex molecules.

Nernst has pointed out that all methods for determining molecular weights in solution involve separation between solvent and solution. In 1890 he devised an elegant method for these determinations, known as the method of the lowering of the solubility, which has received relatively little attention since that time. The method depends upon the degree of lowering of the solubility of two sparingly soluble liquids by adding a solute which is soluble in one of them, but insoluble in the other. Nernst's first experiments were carried out with valeric acid and water as the two sparingly soluble liquids, and the molecular weights of substances soluble in valeric acid, but insoluble in water, were determined by observing the diminution in the titration value of the acid present in the aqueous layer. And just as the lowering of the vapour pressure of a solvent by a dissolved substance is expressed by the equation $\frac{\Delta p}{p} = \frac{n}{N}$, so here the lowering of the solubility is similarly related to the numbers of solute and solvent molecules by the equation $\frac{\Delta s}{s} = \frac{n}{N}$.

In 1896 this method was further studied by Tolloczko who used ether and water as the sparingly soluble liquids and measured the lowering of the solubility by observations on changes of volume when a solute was added.

Determinations of molecular weights in solution by measurements of the lowering of the vapour pressure have received much attention since the early experiments of Raoult, who used a barometric method but found it somewhat inconvenient on account of various practical difficulties. In 1888, Ostwald and Walker devised a method depending upon the losses of solvent suffered by a solution and by the pure solvent when a current of air is aspirated through them. An elegant and ingenious method for determinations on a micro-scale was devised by Barger in 1904. This method depends upon the principle that isotonic solutions have identical vapour pressures. Barger's method consisted in introducing drops of solutions of the solute, the molecular weight of which was being determined, into capillary tubes containing drops of a standard solution of a solute of known molecular weight, the two being separated by a short air space. By observing the two drops with a microscope, provided with a micrometer scale in the eyepiece, it was possible to observe the increase in the size of the one drop and the diminution in the size of the other, according to the relative values of the respective vapour pressures. When neither drop suffered any change in size it was concluded that the solutions were isotonic, and the molecular weight was thus easily found. Coming to more recent times a number of determinations of the vapour pressures of dilute solutions have been made by Frazer and Lovelace who used a sensitive Rayleigh manometer for measuring the differences of pressure. Thus in 1920 they made a careful study of the vapour pressures of aqueous solutions of mannitol and obtained satisfactory results as regards agreement between the observed and the calculated values.

Experiments on the velocity of reaction between substances dissolved in liquids in which there is no obvious reaction with the solvent have shown that the velocity is profoundly affected by the nature of the solvent. Thus, in 1890, Menschutkin studied the rate of formation of tetraethylammonium iodide from triethylamine and ethyl iodide in some twenty different solvents. He found that this reaction takes place very slowly in solvents such as hydrocarbons but very rapidly in ketones and in alcohols.

Thus at 100° C. the rate in benzyl alcohol is nearly one thousand times as great as it is in hexane. Attempts to correlate results of this kind with other properties of the solvent, such as the dielectric constant, have met with practically no success. Still more interesting have been comparisons made between the kinetics of reactions which can take place either in the gaseous or in the dissolved condition. The number of reactions which are available for experiments of this kind is, however, very limited. Few suitable reactions take place under both conditions, and it is obviously important to choose reactions which are truly homogeneous in the gaseous phase. However, in 1923, Hinshelwood and Prichard showed that the thermal decomposition of chlorine monoxide is homogeneous and uninfluenced by the walls of the vessel. In 1931, Moelwyn-Hughes and Hinshelwood showed that the kinetics of this reaction in carbon tetrachloride solution were identical with those in the gaseous condition. Similar results have been obtained for a few other reactions, and it has therefore been concluded that the same principles regarding the discussion of gaseous reactions on the basis of the kinetic theory may be applied to the discussion of reactions in solution. But it by no means necessarily follows that there is any identity in the nature of the two processes. According to Larmor (1897) the parallelism between the laws for gaseous pressure and for the osmotic pressure of dilute solutions, including the numerical identities of the values of the constant R, would seem to depend on the fact that in both cases the molecules are sufficiently far apart to be without appreciable influence upon each other's spheres of action.

Just as the simple gas laws are only valid within limited ranges of pressure, and show considerable departures at high pressures, so the simple laws of osmotic pressure are only true for really dilute solutions. In more concentrated solutions the observed osmotic pressures are much higher than those calculated on the elementary theory. This subject has received careful attention on the experimental side, notably by the Earl of Berkeley and E. G. J. Hartley since 1906, who employed semipermeable membranes of copper ferrocyanide—the substance which has been found to approach most closely to the requirements of an ideal material—and measured the osmotic pressures developed within the apparatus by applying gradually increasing pressure from without, and thus determining the point at which equilibrium between the osmotic and the hydrostatic pressures was established. Most of the experiments were made with cane sugar and with glucose as the solutes, and the apparatus was capable of being used for pressures of upwards of 130 atmospheres. All the results showed very marked departures from the requirements of van't Hoff's theory except for highly dilute solutions. Thus Berkeley and Hartley found that the osmotic pressure of solutions of cane sugar containing two gramme-molecular weights of the solute per litre was more than double that of the theoretical value. These results have been confirmed and extended by a different method by Frazer and Myrik in 1916 and by Frazer and Lotz in 1921, who carried their experiments to much higher pressures. It has been pointed out that the deviations between the observed and calculated values of the osmotic pressure are less pronounced if the concentrations are referred to the *solvent* as distinct from the solution, but they are nevertheless well marked. Attempts to account for these discrepancies by applying the van der Waals equation or some modification of that equation have met with very little success. It has been suggested by some that the abnormally high values of the osmotic pressures may be explained on the basis of some form of combination between the solute and the solvent. Equally unsuccessful have been the numerous attempts to find some explanation of the action of the semi-permeable membranes. Here it may be noted that semi-permeable membranes appear to be selective to some extent. Thus copper ferrocyanide has been found to be very satisfactory for experiments with solutions of sugars, but not for many other substances.

The problems connected with the determination of solubilities of substances have long occupied the attention of chemists, and as far as approximate determinations are concerned they present no particular difficulties, but when accurate values are sought the experimental difficulties are considerable. In the discussions of some of the results which have been published for the solubilities of many well-known substances by different investigators, it must be remarked that the agreement is by no means as good as might reasonably be expected. There can be little doubt that some of the discrepancies are to be traced to the difficulty of attaining complete equilibrium between the solute and the saturated solution. This subject has occupied the attention of some experimenters, notably Lord Berkeley, who in 1904 carried out some very accurate experiments on the solubility of several salts. He found that continuous stirring for a long period, in some cases for upwards of 48 hours, was necessary to attain the saturation value. Another circumstance which exerts an important influence, particularly with very sparingly soluble substances, is the size of the particles. This subject received careful experimental study at the hands of Hulett in 1901. He found that the solubility of gypsum could be increased by nearly 20 per cent by reducing the size of the particles below a certain limit, and in the case of barium sulphate he found that the solubility of the salt having the smallest particles (about 0.2μ in diameter) was nearly double that of the ordinary value. These results are very relevant to the problem of obtaining accurate values for the gravimetric determination of sulphates, an operation by no means free from difficulty. There is a close parallelism between the influence of the size of the particles of a substance on its solubility and on its vapour pressure, just as the vapour pressure of very small drops is greater than that of larger ones. Pawlow in 1909 showed directly that the vapour pressures of compounds such as anthracene, benzophenone, and iodoform was markedly greater when these compounds were in micro-crystalline condition as distinct from a slightly larger crystalline state.

In order to define the condition of a solution as unsaturated, saturated, or supersaturated, all that is necessary is to introduce some crystals of the solute and observe whether any effect is produced or not. As regards supersaturated solutions, a distinction has been drawn by Ostwald between such as require a nucleus of the solid to induce crystallization and those which crystallize spontaneously, the former were termed metastable and the latter unstable or labile systems. The distinction between the metastable and unstable states is not always easy to draw, but as regards solutions, Miers and Miss Isaac in 1906 were able to trace out a curve for various salts which they termed the supersolubility curve situated very nearly parallel to the solubility curve, and represents the temperature and concentration of each solution as it passes from the metastable to the labile condition. This supersolubility curve can be traced out with the aid of observations on the refractive index of the solutions. Between the two curves the condition of the solution is truly metastable and requires nuclei of the solute to bring about crystallization.

Although many determinations of solubilities had been made from early times, it is only since the year 1887 that the general principles regarding the interpretation of solubility curves, particularly of hydrated salts, have been properly understood. This has been directly due to the work of Roozeboom who pointed out the necessity of applying the phase rule to these problems. Complex systems such as the solubility relations of the hydrates of calcium chloride (1889) and of ferric chloride (1892) were worked out by Roozeboom. Indeed, the vast literature regarding solubility which has grown up since that time is simply a collection of practical results which depend directly on Willard Gibbs's thermodynamic equations. Of particular importance have been the various conditions which determine the formation and decomposition of double salts, a subject which has developed since about 1887 associated more particularly with the names of van't Hoff, van Deventer, and Meyerhoffer. One example of the practical value of this kind of work may be quoted by way of illustration. In 1848 Pasteur effected the resolution of a saturated solution of sodium ammonium racemate by allowing it to crystallize at the ordinary temperature, and thus obtained separate crystals of *dextro-* and *laevo*-sodium ammonium tartrates respectively. A few years later the experiment was repeated by Staedel who failed to bring about the resolution of the racemate. The explanation of these results was given by van't Hoff and van Deventer in 1887, who showed that there is a transition temperature of 27° C. below which the stable phases in contact with the saturated solution consist of the *dextro-* and *laevo*-sodium ammonium tartrates, whereas above this temperature the racemate constitutes the stable phase. The general principles regarding the conditions whether a saturated solution of two salts which can form a double salt will deposit crystals of the two constituents or of the double salt has been fully worked out chiefly by van't Hoff, and found to be essentially determined by the respective contents of the water of crystallization of the double salt and of the constituent single salts taken together; because the thermal changes connected with the addition or the removal of water are the most important thermochemical effects which take place, and thus determine the sign of the total heat effect—a result in fundamental agreement with the well-known theorem of Le Chatelier, first formulated in 1884.

The problems connected with the crystallization of substances from solution have occupied the attention of chemists for many years, and success in the interpretation of results has been directly due to the application of the principles of the phase rule. The term *cryohydrate*, which was introduced by Guthrie as long ago as the year 1875 to denote a product of apparently definite composition obtained by crystallizing an aqueous solution of a salt, would seem to suggest a compound of some sort. Later, in 1884, the term *eutectic*, applied by the same author to characterize an alloy of minimum melting-point, was not intended to convey the idea of a definite compound. This idea has been amply confirmed by later work, particularly as alloys characterized as eutectics from a study of their freezing-point diagrams have been shown to possess a heterogeneous micro-crystalline structure. The essentially heterogeneous nature of cryohydrates has also been fully demonstrated, and the term is now of historic interest only, but the term eutectic has remained in general use. In 1890, van't Hoff, when considering certain abnormal results which had been obtained in cryoscopic determination of molecular weights, pointed out that these irregular values were due to some of the solute freezing out together with the solvent. In this way he developed the conception of solid solutions, a subject of particular importance in connexion with the study of alloys. In 1903, Lash Miller and Kenrick applied the phase rule to the study of basic salts. They pointed out that in order to characterize a material of apparently definite composition as a true basic salt, as distinct from a mixture, it is necessary to show that the solid phase remains of constant composition although the composition of the liquid from which it is deposited may be varied over a range.

In 1849, Graham began a series of experiments on the diffusion of dissolved substances which he extended over a number of years, and which may be said to have laid the foundations of the vast subject of colloid chemistry. He found that certain substances, such as acids, bases, and salts, and many other substances of a crystalline character, diffused in solution at rates which could easily be measured, whereas substances of a different type such as gelatine, glue, and silicic acid diffused with such extreme slowness as to be regarded as virtually non-diffusible. The former class of substances were termed crystalloids by Graham, the latter he designated as colloids on account of their gelatinous or non-crystalline character. In 1861, Graham found that remarkably complete separations of the two classes of substances could be effected by a process devised by him and named *dialysis*, which consisted in enclosing the solution within a membrane such as parchment which permitted the passage of the dissolved crystalloid, but resisted that of the colloid. He turned this to practical account in toxicological analysis, as he found that practically all the well-known poisonous substances were of the diffusible type, and therefore capable of being easily separated from organic matter which would give rise to difficulty in analysis.

Although Graham is rightly recognized as the founder of colloidal chemistry as a special department of study, it should be noted that certain substances in a colloidal condition in solution had been recognized previously without a definite name having been given to characterize them. Thus Faraday in 1857 obtained gold in what was most certainly colloidal solution by reducing dilute solutions of auric chloride by means of phosphorus, and obtained liquids of a ruby, violet or blue colour by slight variations in the experimental conditions. Faraday recognized that these solutions consisted of the metal in a state of extremely fine dispersion. At a much earlier date, some two centuries before this time, the material known as purple of Cassius had been obtained by reducing dilute solutions of auric chloride with stannous chloride. This purple of Cassius has been the subject of many investigations. The older generations of chemists regarded it as a compound, and Berzelius actually assigned a formula to it as a double oxide of gold and tin. It is now universally recognized to be a mixture or an adsorption product of colloidal gold and colloidal stannic oxide.

Graham considered that crystalloids and colloids were fundamentally distinct classes of substances, and having regard to the types of compounds with which he conducted his experiments, such a conclusion was a reasonable one at the time. The chief advance which has been made since the time of Graham is that there is no such distinction between two classes of substances, but rather that *any* substance can, by suitable experimental technique, be obtained in colloidal condition. Indeed, the passage of certain very sparingly soluble substances into colloidal solution, a phenomenon noted by various chemists including Graham who termed it *peptization*, while a source of difficulty in gravimetric analysis, has incidentally been a means of preparing such solutions. Two general methods of preparing these sols, as they were first termed by Graham, have been devised, namely, *condensation* methods and *disintegration* methods. Methods which involve starting from substances in true solution reacting so as to produce a sparingly soluble substance as a new phase may be classed as condensation methods. As regards metals these condensation methods are almost universally methods which involve reduction. Thus numerous reducing agents other than phosphorus have been used with success for the preparation of hydrosols of gold. Such reducing agents as formaldehyde, hydrazine, sodium hydrosulphite, and many others have found application in the preparation of colloidal solutions of silver, gold, and of the platinum metals. It will be obvious that such methods must necessarily result in the production of liquids which contain other substances in addition to the metal in the colloidal condition. The removal of these foreign substances may be effected afterwards by some process of dialysis. For many purposes, however, the removal of the products of the reaction is not necessary, and indeed is sometimes actually disadvantageous, as it has been found that certain types of reducing agents, usually organic compounds, increase the stability of the colloidal solutions. Some very interesting experiments were carried out about the year 1886 on the preparation of colloidal solutions of silver by Carey Lea. He found that this metal could be obtained in the colloidal condition by reduction of silver nitrate with a ferrous salt in the presence of sodium citrate. The presence of the salt of the organic acid was found to be necessary to obtain stable sols, and doubtless the reaction was accompanied with the production of some substance which acted as what is now termed a protective colloid.

Disintegration methods or, as they are sometimes termed, dispersion methods of preparing colloidal solutions consist essentially in treating a substance in the solid state in such a manner as to bring it into the sol condition. The process of peptization, to which reference has already been made, in which a sparingly soluble precipitate is overwashed in gravimetric analysis and thus passes through the filter, is a familiar method of producing a colloidal solution of the very sparingly soluble compound. Of great importance for the preparation of metallic sols is the method of electric dispersion devised by Bredig in 1898. This method consists in establishing an electric arc between two stout terminals of the metal under water. A current of 5–10 amperes at a potential of 50 volts was found convenient for preparing hydrosols of metals such as platinum. Bredig made elaborate studies of the properties of colloidal solutions of platinum prepared in this way, and found that the liquid possessed well-marked catalytic properties. When a small quantity of colloidal platinum was introduced into a solution of hydrogen peroxide rapid decomposition of the compound took place. Bredig also found that the catalytic properties of these platinum hydrosols were greatly impeded by the action of poisonous substances, such as hydrogen cyanide, mercuric chloride, or arsenious oxide. The poisoning of the colloidal platinum appears to be due to the adsorption of the poisonous substance by the colloidal particles and thus interfering with their catalytic action. Bredig's original method for preparing these colloidal solutions has been greatly improved, especially by Svedberg since 1905, who has found that much better results were obtained by using alternating currents from an induction coil for producing the electric discharge. In this way Svedberg was able to prepare colloidal solutions of metals of widely different chemical properties including sols of the alkali metals in inert organic liquids.

Of the numerous inorganic substances which have been studied in the colloidal condition, few have been of such interest or importance as regards fundamental discoveries as arsenious sulphide. As long ago as the year 1882, Schulze observed that if arsenious oxide is allowed to react with sulphuretted hydrogen in aqueous solution arsenious sulphide is indeed produced, but is not precipitated. But if hydrochloric acid is added to the liquid the arsenious sulphide, present in colloidal solution, is readily obtained as a precipitate. The presence of certain electrolytes in the solution was thus shown to be essential for the arsenious sulphide to separate in a flocculated condition. These hydrosols of arsenious sulphide have been the subject of study by subsequent investigators, and particularly by Linder and Picton between 1892 and 1905. It was very largely as the result of their experiments, in the course of which they succeeded in preparing four different types of hydrosols of arsenious sulphide, and embodied their results in papers bearing the singularly appropriate title of *solution and pseudo-solution*, that it is now recognized that there is no sharp line of distinction between solutions and suspensions, but rather that there are various intermediate mixtures which cannot be relegated to one or other class. The four types of arsenious sulphide hydrosols which Linder and Picton prepared were obtained by them as follows: α, by the action of hydrogen sulphide upon arsenious oxide dissolved in potassium bitartrate, followed by displacement of the excess of hydrogen sulphide by hydrogen; β, by the action of hydrogen sulphide upon a solution of an alkali arsenite, followed by dialysis; γ, from a concentrated solution of arsenious oxide by the action of hydrogen sulphide, followed by removal of the excess of sulphuretted hydrogen; δ, as in the last-named but starting from a very dilute solution of arsenious oxide. All these solutions showed the Tyndall effect, but in other respects there were important differences between them. The α sol showed actual particles which were visible under the microscope; the β sol showed no particles which could be seen with the microscope, but there was no evidence of diffusion in the solutions; the γ sol showed diffusion, but when the liquid was filtered through unglazed earthenware some of the particles were removed from solution; and lastly the δ sol was readily diffusible and could be passed through unglazed earthenware without suffering any change.

An important advance in the study of colloidal solutions was made in 1903 by Siedentopf and Zsigmondy by their invention of the ultramicroscope. This apparatus consisted of a device for rendering visible particles having dimensions considerably below the visual limits of an ordinary microscope, by causing them to render visible light diffracted from the actual colloidal particles. The ultramicroscope does not thus enable the observer to see the colloidal particles, but to become aware of their existence in a liquid by seeing the light scattered by them. With this apparatus the fundamental work of Linder and Picton has been verified and extended, and in the hands of Zsigmondy it has been of great value in the study of colloidal solutions of gold, particularly as regards the relation between the colours of the solutions and the size of the particles.

The experiments of a number of investigators have shown that the particles in colloidal solutions are electrically charged, and that the presence of the charge is in some way necessary for the stability of the sol. Removal of the charge results in precipitation. This subject has been approached from two different angles, namely, by observing the behaviour of colloidal solutions when a pair of electrodes maintained at a high difference of potential is immersed in the liquid, and by studying the effects of causing flocculation to take place by adding electrolytes. The former method, formerly known as electric endosmose and later as electrophoresis, was employed by Linder and Picton in their investigations on colloidal arsenious sulphide from which they concluded that the particles of that substance were negatively charged. They also found that solutions of colloidal ferric hydroxide, prepared from solutions of ferric chloride by hydrolysis and removal of the resulting hydrochloric acid by dialysis, were positively charged because the particles wandered in the opposite direction to those of arsenious sulphide. These conclusions have been verified and extended by studies on the flocculating action of solutions of electrolytes on these colloidal solutions. The early experiments of Schulze on arsenious sulphide may be said to have established an important relation between the valency of an ion and its capacity for causing flocculation of the colloidal particles. Schulze in 1882 found that the relative coagulative powers of univalent, bivalent, and tervalent ions were approximately in the ratios 1 : 30 : 1650. Very similar figures were obtained in 1895 by Linder and Picton for sols of arsenious and antimonious sulphides. It is thus evident that the capacity of an ion for effecting flocculation is some function of its valency. Further confirmation of this principle was given by Hardy in 1900, who pointed out that the coagulative properties of any salt are determined by the valency of one of its ions and that the active ion is always of opposite electrical sign to that on the colloidal particles. An interesting attempt to explain these remarkable phenomena made by Whetham in 1899, on the basis of the minimum electrostatic charge necessary to cause a sufficient aggregation of colloidal particles to induce actual flocculation, resulted in expressing the coagulating powers of equivalent solutions containing univalent, bivalent, and tervalent ions as proportional to $1 : x : x^2$, where x is an integer. The results were in fair agreement with a formula of this kind, but later work has shown that it cannot be regarded as more than an approximation, since it has become clear that the valency of an ion is not the sole factor which determines its coagulative capacity, but that its specific properties are not without influence. In particular it seems to be well established that the hydrogen ion, or more correctly according to modern usage, the hydroxonium ion, is considerably more efficient as a flocculating agent than other univalent ions.

For many years it was accepted that the sign of the charge on colloidal particles was determined by the nature of the substance. Thus ferric hydroxide was described as a positive colloid, whereas the sulphides of arsenic and antimony were classified as negative colloids. When colloidal solutions containing oppositely charged particles were mixed together mutual flocculation was found to take place as the result of neutralization of the charges. Further studies on this subject, particularly experiments by Lottermoser in 1905 on sols of the silver halides, have shown that the sign of the charge on colloidal properties is determined, not by the nature of the substance, but by the way in which the colloidal solution is prepared. The classification of ferric hydroxide as a positive colloid and of arsenious sulphide as a negative colloid arose because these solutions were always prepared in such a way as to result in the particles having their characteristic electric charges. It was shown by Lotter-moser that when highly dilute solutions of silver nitrate and potassium iodide were mixed together the resulting hydrosol of silver iodide was positively charged if the silver nitrate was in excess, whereas an excess of potassium iodide resulted in the hydrosol being negatively charged. It would thus appear that the sign of the charge on the colloidal particles of silver iodide is determined by the adsorption of silver or of iodide ions. When silver ions are adsorbed the particles are charged positively, while a negative charge is assumed when the particles adsorb iodide ions.

The capacity of the silver halides to adsorb either cations or anions has had practical developments since 1923 when an important advance was made in volumetric analysis by Fajans who introduced some indicators of a type known as adsorption indicators for titration work with silver nitrate. The introduction of adsorption indicators into volumetric analysis may truly be claimed as one of the most important advances in this branch of practical chemistry since potassium chromate was first used as an indicator by Mohr in 1856 for work in neutral solution, or since 1874 when Volhard devised his well-known thiocyanate method for titrating silver in acid solution. Fajans found that fluorescein and certain halogenated fluoresceins, such as eosin (tetrabromofluorescein) and Rose Bengal (dichlorotetraiodo-fluorescein), undergo striking colour changes at the end-point when solutions of halides are titrated with silver nitrate. The colour changes with such dyestuffs take place upon the colloidally dispersed particles or upon the surface of the precipitate, in consequence of the adsorbed silver ions combining with the anions of the dye to form a compound, or alternatively, the adsorbed halide ions may combine with the cations of the dye-stuff. These fluorescein dyestuffs are practically restricted to titration work in neutral solution, but other compounds, particularly tartrazine and phenosafranine, have been found to yield excellent results in acid solutions.

Since the silver halides differ very greatly in their degree of insolubility and in their capacity for adsorbing the ions of various dyestuffs, methods have been devised for determining an iodide and a chloride in the same solution, and with an accuracy of some three parts per thousand. Fajans has discussed the colour changes which take place with these adsorption indicators in terms of ionic deformation phenomena on the crystal lattice of the silver halide, but alternative explanations based on tautomeric changes of the dyestuffs are also possible.

When a very sparingly soluble compound is produced by double decomposition the character of the precipitate may depend very considerably upon the conditions under which it is produced. Most writers on gravimetric analysis have given careful directions for carrying out the precipitation of substances used in determinations so as to secure that the compounds shall separate in as pure condition as may be possible. This subject was studied in a very interesting way by von Weimarn between 1911 and 1914. He defined a dispersion coefficient to give expression to a relation between the concentration of an 'insoluble' substance before actual precipitation to the limiting value of the solubility, and considered that it is the degree of supersaturation of the solution which determines the physical character of the precipitate. Von Weimarn was able to support his ideas by some interesting experiments on the precipitation of barium sulphate by mixing *equal* volumes of *equivalent* solutions of barium thiocyanate and manganous sulphate in concentrations ranging from $7N$ down to $N/20,000$. The degree of supersaturation was thus varied over an enormous range, and it was found that the smaller the degree of supersaturation the larger were the resulting crystals of barium sulphate. At very high degrees of dispersion the precipitates were always amorphous and even gelatinous.

All the experimental evidence which has been accumulated goes to show that the distinction between the crystalline and amorphous states of matter, just like that between true solutions and colloidal solutions, is very difficult to define. A colloidal substance may or may not be amorphous. Thus Scherrer in 1918 was able to show by X-ray analysis that the ultramicroscopic particles in gold hydrosols are actually crystalline. When a substance is spoken of as being in a colloidal condition, the emphasis is directed upon the liquid being a disperse system. The dispersed substance within the dispersion medium may be a solid, as is the case with metals, metallic sulphides, and hydroxides, and other compounds of a similar character, or it may be a liquid as with emulsions. This leads on to the somewhat difficult problem of the classification of colloidal solutions. It will have been noted that the term colloid is now used in a considerably wider sense than ever it was understood by Graham. An attempt to classify colloids as *suspensoids* and *emulsoids*, according as the disperse phase is a solid or a liquid, was made by Wolfgang Ostwald about the year 1907. Some five years later a classification into *irreversible* or *reversible* colloids was suggested by Zsigmondy. It is well known that colloidal solutions of the type classified as suspensoids when once separated from solution cannot in general be redissolved, whereas numerous colloids of a gelatinous character can be brought back into solution. Zsigmondy's classification, in short, is based upon the reversibility or otherwise of the transition from the gel to the sol condition. Such a classification, though at first sight apparently simple, is nevertheless open to serious objections on practical grounds, since the stability of many inorganic colloids, such as metals, is profoundly increased by the addition of substances of the type of albumen or gelatine which act as protective colloids. A system of classification into *hydrophobic* and *hydrophilic* colloids, first suggested by Perrin in 1905, has come into general use, largely owing to the influence of Freundlich. This system corresponds fairly closely to that of suspensoids and emulsoids, but has the advantage of not being in any way committed as to whether the dispersed substance is a solid or a liquid.

The lyophilic colloids are in many respects very different from the lyophobic sols. Thus the lyophilic sols are much less sensitive to the action of electrolytes, the viscosity of the solutions is considerably greater than that of the dispersion medium, and experiments on their osmotic properties would appear to indicate that there is a perceptibly closer approach between them and true solutions than is the case with the lyophobic sols. The transition from the sol to the gel condition has been the subject of numerous investigations with these lyophilic colloids, and the results obtained have in many instances shown how very difficult it is to decide whether a product to which a formula may be assigned is or is not a definite compound. The same difficulty regarding the problem as to whether certain materials should be considered as compounds or not has frequently arisen in connexion with the study of adsorption phenomena. It will be seen in what follows that instances have arisen in which supposedly well-defined compounds have later been regarded as adsorption products, but the difficulty has not always been ended at that stage, as further work has in some cases shown that complications have arisen regarding the interpretations to be put upon the experimental work. One or two examples may be quoted by way of illustration.

In 1834, Bunsen discovered that freshly precipitated ferric hydroxide was a very efficient antidote for poisoning by arsenious oxide, and he considered that its action in this respect was due to the formation of an insoluble basic ferric arsenite. This conclusion was challenged by the work of Biltz in 1904, who carried out a number of experiments on the removal of arsenious oxide from solution by ferric hydroxide, and he found that the quantitative relations indicated clearly that no compound was formed, but on the contrary the arsenious oxide was removed solely by adsorption. Although some further experiments have been carried out in more recent years on this subject, the results of Biltz seem to be generally upheld.

The well-known blue product obtained by the action of iodine upon starch has long been used as a means of detecting the element, and of enabling the end-point in iodometric titrations to be determined. The older generation of chemists was strongly of opinion that the blue product was some compound of starch and iodine. Somewhat later it became recognized that a trace of an iodide appeared to be necessary for the production of the blue colour. A great many investigations have been carried out on this material, but particular interest is attached to the experiments of Küster about 1894, who studied the distribution of iodine between starch and water and concluded that an adsorption product was formed. It may be noted that starch is by no means the only substance which is turned a blue colour by iodine. Thus basic lanthanum acetate produces a blue colour with this halogen, but the formation of this colour has not been found satisfactory as a test for that element, as certain other earths behave in a similar manner. Coming to more modern work on the starch-iodide product, Barger and Miss Field in 1912 showed that certain other organic compounds, e.g. saponarin, as well as starch, could yield blue materials. They concluded that in order to obtain the blue colours it was necessary to have the adsorbing substance present in the colloidal condition, but not in true solution. Barger afterwards found it very difficult to come to a definite conclusion regarding the nature of the starch-iodide products, and pointed out that while investigations extending over a period of forty years had amply demonstrated that iodine is adsorbed by starch from solution in varying proportions depending upon the composition of the solution, it does not follow that the possibility of formation of a compound is necessarily ruled out. He pointed out that it is not simply a question of 'adsorption or chemical union', but rather that the iodine may be taken up by the starch by both processes. There is nothing to prevent a compound containing a definite proportion of iodine from adsorbing more iodine from solution according to the conditions which may prevail. Bancroft (1932) commenting upon this remarked that 'Barger hedges unnecessarily'.

The literature of inorganic chemistry abounds with substances, described as compounds, the existence of which is most certainly doubtful. Thus while there is no doubt that the silicates are definite compounds, it is very questionable whether silicic acid can be so described. It is a simple matter to obtain silica containing water in such proportions as to correspond with the formula of more than one silicic acid, but the water may not necessarily be present in the combined condition. Numerous experiments on the nature of amorphous and gelatinous hydroxides were carried out by van Bemmelen during some twenty years since 1888. His object was to ascertain whether the water in these substances was present in the combined or in the adsorbed condition, by determining the pressure-concentration relations so as to find out whether the pressure of the water vapour was such as to be consistent with univariant systems, and thus indicate the presence of compounds according to the phase rule. Van Bemmelen's method was to employ a number of desiccators containing sulphuric acid of different concentrations so as to provide atmospheres of known vapour pressure, and to introduce the substances into these atmospheres and weigh them after equilibrium had been considered to have been attained. The results were very complicated and in many respects difficult to interpret. The curves of isothermal dehydration did not show abrupt discontinuities, nor were there any discontinuities when the substances were rehydrated. But the curves of withdrawal and of addition of water were not coincident, and hysteresis was well marked. The general conclusion to be drawn from these experiments was that as regards gelatinous or amorphous substances the water is adsorbed, but is not chemically hydrated. For a number of years van Bemmelen's conclusions were accepted without question, but gradually doubts began to arise regarding the interpretations to be put upon his curves. The question at issue was whether equilibrium was really attained or not. If equilibrium was not completely established the conclusions might be altogether in error. The subject has been resumed in more recent years, notably by Willstätter and his collaborators since 1923, who have examined the effect of withdrawing water from inorganic hydroxides by intensely hygroscopic agents, such as anhydrous acetone. Their general conclusion seems to have been that in *crystalline* hydroxides, the water is really chemically hydrated, whereas in amorphous or gelatinous materials the water is adsorbed.

It has been remarked that while the older chemistry was concerned mainly with substances, modern chemistry is much more concerned with solutions. The problems connected with the study of osmotic phenomena and of colloids provide abundant demonstration of the truth of this statement. Even at the present time, although the investigations date back well into the nineteenth century, the nature of the processes of osmosis and dialysis are, at best, imperfectly understood. A semipermeable membrane permits the passage of pure solvent, but resists the passage of substances in true solution. A membrane which acts as a dialyser effects the separation of crystalloids from colloids in solution. The process devised by Bechhold in 1907, known as ultrafiltration, in which membranes of collodion or of gelatine hardened by formaldehyde are used, enables colloidal particles to be removed from solution. Bancroft in 1925 expressed the view that the difference between an ultra-filter and a semi-permeable membrane is by no means unimportant if a fundamental distinction between a true solution and a colloidal solution is to be drawn.

REFERENCES

J. H. VAN'T HOFF. *The Foundations of the Theory of Dilute Solutions.* Alembic Club Reprints, No. 19.

J. H. VAN'T HOFF. *Lectures on Theoretical and Physical Chemistry.* Three volumes. English edition. London, 1898–1900.

T. GRAHAM. *Chemical and Physical Researches.* Edinburgh, 1876.

J. LARMOR. *Aether and Matter.* Cambridge, 1900.

A. FINDLAY. *Osmotic Pressure.* 1913; second edition, 1919.

A. FINDLAY. *The Phase Rule.* 1904; eighth edition, 1938.

H. FREUNDLICH. *Kapillarchemie.* 1909; second edition, 1922. English translation of the second edition. London, 1926.

E. HATSCHEK. *The Foundations of Colloid Chemistry.* A collection of Early Papers bearing on the subject. London, 1925.

P. P. VON WEIMARN. *Zur Lehre von den Zuständen der Materie.* Dresden und Leipzig, 1914.

J. M. VAN BEMMELEN. *Die Absorption.* Dresden, 1910.

TH. SVEDBERG. *Colloid Chemistry.* American Chemical Society Monograph Series, No. 16, 1928.

WILDER D. BANCROFT. *Applied Colloid Chemistry. General Theory.* New York, 1932.

K. FAJANS. *Adsorption Indicators for Precipitation Titrations.* Part VII of *Newer Methods of Volumetric Analysis.* Edited by W. Böttger and translated by R. E. Oesper. U.S.A. printed. London, 1938.

Annual Reports of the Chemical Society since 1904.

Chapter VIII

SOME ESSENTIAL FEATURES OF CHEMICAL CHANGE

Henri Poincaré in *La Science et l'Hypothèse* has called attention to two opposite tendencies which have characterized the development of physics, viz. the tendency towards unity and simplicity whereby new discoveries are assimilated into the existing structure, and thus tend towards greater clarity in the whole science, and the tendency towards diversity and complexity whereby phenomena considered to be simple are not really so when investigated by more refined experimental methods. The history of physics has shown that each of these tendencies seems to have triumphed in turn. Very similar has been the history of chemistry, and Poincaré's remark, 'C'est un malheur pour une science de prendre naissance trop tard, quand les moyens d'observation sont devenus trop parfaits. C'est ce qui arrive aujourd'hui à la physico-chimie; ses fondateurs sont gênés dans leur aperçus par la troisième et la quatrième décimales', seems to be relevant to some branches of the subject. General principles laid down as the result of scanty or even inexact experimental work have sometimes been established on a firm foundation when subjected to the rigid scrutiny of exact measurement. Thus the law of the conservation of mass, first clearly enunciated by Lavoisier as the result of experiments on vinous fermentation—a very crude experimental foundation—was subjected to a very rigorous test by Landolt, Heydweiller, and by others between 1901 and 1908, who examined various inorganic reactions with every possible refinement for carrying out the weighings. The final result of Landolt's last experiments was a verification of the law of the conservation of mass within the limit of accuracy of one part in ten million.

More frequently, however, it has been found that phenomena which at first appeared simple, have on more refined study proved to be complicated. The apparent simplicity has resulted in the laws thus formulated becoming approximations, but nevertheless of considerable value. Thus the simple gas laws are by no means exact but sufficiently close approximations for many purposes. For greater degrees of exactness recourse must be had to the van der Waals equation or to one of its modifications. The study of chemical change has provided numerous examples of reactions, at one time considered to be fairly simple, but which have been found on minute investigation to be more or less complicated.

It has been remarked on more than one occasion that whereas the main trend of nineteenth-century chemistry was directed towards investigating the *products* of chemical change, that of more modern chemistry was much more concerned with studying the *course* of chemical reactions. This is certainly true, especially if the comparison is made by some sort of numerical estimate of the volume of published work on preparative chemistry as compared with that on chemical dynamics. But it is not by any means wise to attempt to draw conclusions regarding the progress in any particular department of chemical science by estimates based upon the amount of published work associated with it. It has sometimes happened that investigations on preparative and analytical chemistry have given rise to problems involving the methods of physical chemistry. The history of the law of mass action between the publication of Berthollet's *Essai de Statique Chimique* in 1803 up to the years 1864–7 associated with the researches of Guldberg and Waage affords an interesting example of how a theorem of fundamental importance in chemistry has been gradually established, not by any means solely as the result of the work of theoretical and physical chemists, but rather in consequence of co-ordination of their work with that of others, who have striven with varying degrees of success to advance the science in altogether different directions.

Even before the publication of Berthollet's work, some small beginnings relating to the influence of the relative proportions of reacting substances on the course of chemical change are to be found in the writings of Bergman (1775) and of Wenzel (1777). Berthollet's publication was, however, of particular importance, even although he was in some respects seriously in error as he ventured the opinion that the composition of compounds was dependent upon the experimental conditions of their formation. In this he was resolutely opposed by Proust, who was able to establish the principle of constancy of composition. It would appear that the controversy centred round the reversibility of chemical reactions; indeed, the two men were to some extent at cross-purposes.

Since the time of Berthollet, problems in chemical dynamics have been attacked by two different types of methods, namely, by kinetic methods which involve measurements on the rate of chemical transformation, and by static methods which involve investigation of chemical equilibria. The law of mass action, first clearly formulated by Guldberg and Waage, was laid down as the result of both of these types of investigation. The first successful experiments on the rate of chemical transformation were carried out in 1850 by Wilhelmy, who measured the rate of hydrolysis of cane sugar under the influence of acids. The course of the reaction was followed by polarimetric measurements, the cane sugar being dextro-rotatory and the products of the reaction, equimolecular quantities of glucose and fructose, having a slight laevo-rotation. Wilhelmy was able to show that the rate of transformation of the cane sugar into invert sugar was proportional to the amount of cane sugar present at any instant, and that it underwent transformation according to a logarithmic law. Wilhelmy has been rightly described as the founder of chemical kinetics. Some very important work was carried out by Harcourt and Esson in 1865 and 1866 on velocity of reaction. They first studied the oxidation of oxalic acid by potassium permanganate in dilute sulphuric acid solution, and made very elaborate experiments on the effect of varying the relative proportions of the reactants. The course of the reaction was followed by 'quenching' it with excess of potassium iodide, and thus stopping the oxidizing action of the permanganate, the liberated iodine being determined with a solution of sodium thiosulphate. They found that when the oxalic acid was present in great excess the rate of reduction of the permanganate was proportional to the concentration of the oxidizing agent. They also observed the interesting phenomenon of autocatalysis in this reaction, viz. catalysis of the main reaction by one of the products, in this case by the manganous sulphate formed from the reduction of the potassium permanganate. Harcourt and Esson then studied the reaction between hydrogen peroxide and hydriodic acid. They brought together hydrogen peroxide, potassium iodide, and dilute sulphuric acid in a large volume of water, and devised an ingenious method for maintaining the hydriodic acid at a nearly constant concentration by adding small quantities of sodium thiosulphate as soon as the presence of free iodine was indicated by starch. Although there were some minor irregularities due to side reactions, Harcourt and Esson were able to observe that when the concentration of the hydriodic acid was kept constant, the rate of disappearance of the hydrogen peroxide was very nearly proportional to the amount present at any instant and followed a logarithmic law. These experiments contributed in a very direct way to the establishment of the law of mass action, and indeed Harcourt and Esson may be claimed as independent discoverers of this law.

Investigations of chemical equilibria in homogeneous systems have played a very important part in the development of the law of mass action, and it is practically impossible to say whether the formulation of the law arose more as the result of kinetic or of static experimental work. As regards the study of equilibria, the experiments of Berthelot and Péan de Saint-Gilles on esterification carried out in 1862 and 1863 were of exceptional importance. They investigated the equilibrium between ethyl alcohol, acetic acid, ethyl acetate and water in great detail, and were able to determine the numerical value of what is now known as the equilibrium constant, and to show that its value was the same from whichever side equilibrium was approached. In any case their work was well known to Guldberg and Waage who took it into account in formulating their law. One of the most valuable contributions which they made was a clear statement of the meaning of active mass as molecular concentration; but it is somewhat remarkable that the treatise of the Scandinavian investigators lay neglected, and it so happened that the law of mass action was rediscovered independently by others. Thus Jellett between 1873 and 1875 investigated the distribution of hydrochloric acid between alkaloids in alcoholic solution by observations on the optical rotatory power. He first investigated the distribution of the acid between quinine and codeine when present in quantity sufficient to saturate one alkaloid but not both. In this way he found that the value of the equilibrium constant for the ratio quinine hydrochloride/codeine hydrochloride was 2·03. Precisely similar experiments with codeine and brucine and with brucine and quinine gave values for the ratio codeine hydrochloride/brucine hydrochloride of 1·58, and for the ratio brucine hydrochloride/quinine hydrochloride of 0·32. Jellett pointed out that the product of the three equilibrium constants was very nearly unity; actually the value was 1·026. Van't Hoff drew attention to the importance of this result, and pointed out the general principle that the product of the values of an equilibrium constant in a cycle of chemical equilibria is equal to unity.

Jellett's experiments will bear comparison with those of Thomsen on the distribution of acids between bases in aqueous solution as determined by calorimetric methods. In 1869 he was able to obtain values for what have been termed the avidities of acids by experiments on the heats of neutralization. Thus Thomsen found that when one equivalent of a strong base such as sodium hydroxide was added to a dilute solution containing one equivalent of hydrochloric acid and one equivalent of sulphuric acid, the hydrochloric acid absorbed approximately two-thirds and the sulphuric acid absorbed one-third of the base. From this he concluded that if the avidity of hydrochloric acid be taken as unity, that of sulphuric acid is 0·5, or in other words that hydrochloric acid is twice as strong as sulphuric acid. Thomsen determined the avidities of a number of acids in this way, and thus obtained for the first time a list of the acids arranged in the correct order of their relative strengths. Previous to Thomsen's work there was no satisfactory measure of the strengths of acids, simply because such results as had been obtained were vitiated by the separation of some of the products of the reaction by volatilization, or by the production of insoluble substances, in short because the conditions for the law of mass action to be applicable were not observed. Thomsen's experiments were followed in 1877 by Ostwald's volume-chemical studies on acids. In these experiments Ostwald determined the changes of volume which take place on the neutralization of acids by bases in dilute solution. These experiments were performed by careful determinations of the specific gravities of the solutions, using pyknometers, before and after reaction. The results were then expressed in terms of volume changes, and it is interesting to note that the order of the acids as determined in this way corresponded, with one or two unimportant exceptions, with the order of the avidities as determined by Thomsen's thermochemical experiments. In this way Ostwald was able to assign a definite meaning to the word *affinity*, a word which had frequently been used long before but in a thoroughly illogical manner, as essentially identical with Thomsen's conception of avidity to denote the combining activity of acids for bases, when the comparisons are made under the conditions required by the law of mass action. The soundness of this reasoning has been thoroughly recognized, and it may be noted that other methods for determining the relative strengths of acids, such as by measurements of their catalytic action on reactions the velocity of which is measurable, have given results in fundamental agreement with these statical methods.

Although different methods for comparing the relative strengths of acids give rise to a satisfactory degree of agreement, other properties of acids have to be taken into consideration in the discussion of some of their reactions. Thus the action of nitric acid on metals was a problem of considerable difficulty to more than one generation of chemists. The earlier workers, e.g. Acworth and Armstrong in 1877, were mainly concerned with studying the nature of the gaseous reduction products obtained with different metals and with different concentrations of the acid. Divers in 1883 drew attention to differences in the nature of the reduction products obtained by the action of such metals as zinc, magnesium, and iron, which readily evolve hydrogen from hydrochloric acid, and of metals such as copper, silver, and mercury which do not normally yield hydrogen from hydrochloric acid. For many years it was widely held that the primary product of the action of a metal on nitric acid was hydrogen, which, however, did not appear as such, but was oxidized by the excess of nitric acid to various products, the nature of which was determined by the particular metal, the concentration of the acid, the temperature, and the presence of other substances. There was something to be said for this point of view, since hydrogen can be collected when magnesium is allowed to react with very dilute nitric acid, and the oxidizing properties of nitric acid are closely connected with its concentration, being feeble with very dilute acid. An alternative theory regarded the metal itself as the reducing agent, the reduction taking place directly without the production of hydrogen. It is curious that what may be termed the hydrogen theory was so widely held when it is borne in mind that hydrochloric acid, which is equal in strength to nitric acid in the physico-chemical sense, is without action upon metals such as bismuth, silver, and mercury, which readily dissolve in nitric acid.

An important observation made by Veley in 1889, which indeed gave the clue to modern views on this subject, lay neglected for many years. Veley showed that when copper is revolved in nitric acid the metal dissolves much more slowly than when it is stationary. The cause of this phenomenon is the removal of nitrous acid, which acts as an autocatalyst from the surface of the dissolving metal. Veley showed that if nitric acid is carefully freed from traces of nitrous acid it is almost without action upon metals which dissolve in it at once under ordinary conditions. The subject was reviewed comprehensively by Bancroft in 1924, and in 1926 some interesting experiments were carried out by Hedges, who investigated the effects of revolving the metals and the effects of the presence or absence of nitrous acid. He found that metals which come below hydrogen in the electrochemical series, such as copper and silver, when rotated in nitric acid do not dissolve, because the nitrous acid produced by reduction is not allowed to accumulate at the surface of the metal. Direct addition of nitrous acid to the solution caused the revolving metal to dissolve rapidly. On the other hand, metals such as zinc and magnesium, which appear above hydrogen in the electrochemical series, dissolve rapidly in dilute nitric acid, whether the metal is rotated or not; actually the effect of rotation is to *increase* the rate of dissolution, and nitrous acid has no accelerating effect. The passivity of some metals in nitric acid is due to the production of a film which protects the metal from further action of the reagent. The passivity of metals has formed the subject of many investigations, of which those of Evans since 1927 are of particular importance.

The action of metals on acids is sometimes much more complicated than might be expected having regard to their position with respect to hydrogen and to each other in the electrochemical series. Thus copper can be dissolved by boiling hydrobromic or hydriodic acids with evolution of hydrogen, and if the experiment is carried out quantitatively it will be found that the volume of hydrogen evolved corresponds to the cuprous equivalent of the metal. Even boiling concentrated hydrochloric acid has a similar action upon finely divided copper. The cause of this action is due to the formation of complex halogen anions of the type CuX_n^-. It will thus be evident that the action of an acid upon a metal is determined not only by the tendency to discharge hydrogen ions, but to the capacity for forming complex anions. It is on this account that many metallic oxides dissolve much more readily in hydrochloric acid than they do in nitric acid—a phenomenon well known to analytical chemists—although both acids are of practically identical strength. Similar considerations are applicable to the solvent action of potassium cyanide on many metals. Thus copper dissolves in aqueous potassium cyanide with evolution of hydrogen and formation of a solution of potassium cuprocyanide. Faraday discovered that finely divided gold is dissolved by dilute solutions of potassium cyanide, and in 1885 MacArthur and Forrest applied this discovery technically in gold extraction. The presence of atmospheric oxygen was found to be necessary for the production of potassium aurocyanide, and this subject was investigated fully by Bodländer in 1896, who showed that autoxidation is involved.

Few reactions have been of such importance in the subsequent developments of chemical dynamics, and indeed of certain other branches of chemistry, as the hydrolysis of cane sugar under the influence of acids, the reaction first studied by Wilhelmy in 1850. In the hands of Ostwald and his pupils it has provided a means for comparing the relative strengths of acids by a kinetic method, and thus confirming and extending the results obtained by the earlier static methods. Since 1889, Ostwald and others conducted numerous investigations on the velocity of reactions which are catalyzed by acids, and supplemented them with measurements of the electrical conductivities of the acids. These experiments had much to do with the development of the classical theory of electrolytic dissociation, but it would be right to say that so far as kinetic methods for studying the strengths of acids are concerned, the hydrolysis of cane sugar was the predominant reaction to receive attention. In this connexion it may be noted that while the catalytic method resulted in placing the acids in the same order of strength as judged by the older statical methods and by the comparison of the electric conductivity first devised by Arrhenius; experiments on the effect of neutral salts on the catalytic action of acids were among the first steps which led to the gradual recognition of the fact that the properties of an acid are not determined *solely* by the concentration of the hydrogen ions derived from it.

In 1890 a most important investigation on the hydrolysis of cane sugar into glucose and fructose under the influence of the enzyme invertase was carried out by O'Sullivan and Tompson. They were able to show that the rate of inversion of cane sugar, under the influence of this enzyme, could always be expressed by a time curve of the type described by Harcourt and Esson characteristic of a chemical change of which 'no condition varies excepting the diminution of the changing substance'. The effect of rise of temperature upon the velocity of reaction, viz. an approximate doubling of the rate of hydrolysis for a rise of 10° C., was found to be closely similar to what had been observed for other reactions until a temperature of about 55–60° C. was reached. Above this temperature the invertase was gradually destroyed, and at 75° C. destruction was complete. O'Sullivan and Tompson at first encountered considerable difficulty in following this reaction with the aid of a polarimeter, as they found that the dextrose formed by the action of the invertase was initially 'in the birotary state', and consequently the optical activity of the solution was not a reliable guide to the amount of inversion which had taken place, but they were able to obviate this difficulty by adding caustic alkali to stop the action of the enzyme and allow sufficient time for the rotation to assume a steady value. It was definitely established as the result of this investigation that a chemical reaction which takes place under the influence of an enzyme follows the same laws as if it took place under the influence of any other catalyst, so long as the experiments are conducted within the limits of temperature consistent with the stability of the enzyme. Although some of the conclusions of O'Sullivan and Tompson were challenged by subsequent investigators, their essential accuracy was confirmed in 1908 by Hudson, who incidentally verified the result that the hydrolysis of cane sugar by invertase is a reaction of the first order.

Previous to 1898 reactions involving the action of enzymes were essentially examples of reactions involving the degradation of complex molecules into simpler ones, but in that year van't Hoff ventured the suggestion that enzymes might be effective in bringing about syntheses, and an experimental realization of this was accomplished by Croft Hill. It is well known that maltose under the influence of maltase undergoes hydrolysis, the products of the reaction being two molecular proportions of glucose. In 1898, Croft Hill subjected fairly concentrated solutions of glucose to the action of yeast maltase, and he found that the properties of the resulting solutions were such as to indicate the formation of a certain amount of maltose, thus showing reversal of a well-known reaction. More detailed investigation showed, however, that an isomeric sugar, known as isomaltose, is actually produced under these conditions. Since that time the general problem of the reversibility of enzyme action has been attacked by numerous investigators, and it is now generally admitted that enzyme reactions do not differ from other reactions in any fundamental respects.

In the years following the enunciation of the law of mass action many experiments were made on equilibria in gaseous systems, which produced results, not merely of intrinsic value, but of considerable influence in securing more general recognition of the importance of this law. Thus in 1877, Lemoine carried out a most elaborate investigation on the formation and dissociation of hydrogen iodide. He found that the final condition of equilibrium was identical from whichever side it was approached, and was able to establish certain principles which are generally applicable to reversible gaseous reactions. Thus it was found that rise of temperature had an enormous influence upon the time required for the establishment of equilibrium: in particular this could be counted in *months* at 265° C., in *days* at 350° C., and in *hours* at 440° C. Lemoine also found that the product of the partial pressures of the iodine vapour and of the hydrogen divided by the square of the partial pressure of the hydrogen iodide was approximately constant, as it should be in accordance with the law of mass action. He also found that increase or diminution of the external pressure was without effect upon the equilibrium as it should be, since the reaction takes place without any change of the number of molecules. Twenty years later this work was repeated with improved experimental technique by Bodenstein, who was able to confirm all the fundamental principles which were established by Lemoine.

From the earliest times the effect of rise of temperature on chemical change has attracted attention, but the scientific study of this subject cannot be dated earlier than the latter part of the nineteenth century. Two distinct effects of temperature on chemical change have to be considered, namely, the effect upon chemical equilibrium and the effect upon velocity of reaction. The effect of change of temperature upon equilibrium seems to have been first clearly enunciated by van't Hoff about 1884 in his principle of mobile equilibrium which is really a more particular statement of Le Chatelier's rule, the latter referring to pressure as well as temperature. To Arrhenius belongs the credit of having first formulated a quantitative expression in 1889 for the effect of rise of temperature on velocity of reaction. Having seen from his own experiments and from those of others that in general a rise of 10° C. doubles or trebles the reaction velocity, he derived the equation $\dfrac{d \log k}{dT} = \dfrac{A}{RT^2}$, in which k is the velocity constant of the reaction, T is the absolute temperature, and A is a thermal quantity. This equation is closely similar to van't Hoff's equation $\dfrac{d \log K}{dT} = \dfrac{Q}{RT^2}$, which deals with the effect of temperature on the equilibrium constant K of a balanced reaction, where Q is the heat of the reaction, the constant K being the ratio of the velocity constants of the direct and of the reverse reactions. In the Arrhenius equation the quantity A can be measured in calories per gramme-molecule if R is taken as 2 calories. The accuracy of the equation can be tested by plotting the logarithm of the velocity constant against the reciprocal of the absolute temperature, when a straight line is obtained. It is very remarkable that Arrhenius interpreted his equation by assuming that a certain fraction of the total number of molecules became endowed with additional energy, and were thereby activated and thus enabled to react, the active molecules being formed endothermically from the inactive ones. It will be seen that this view of the activation of molecules is fundamentally correct.

While the study of dissociation phenomena occupied the attention of chemists in the latter part of the nineteenth century as regards the establishment of the fundamental principles, the twentieth century has not only seen the further development of chemical dynamics, but of some of the fruits of the application of the principles of the subject to technical problems.

Some of the most interesting illustrations of the application of physico-chemical principles relating to the effects of physical conditions on gaseous reactions are to be seen in the methods of bringing atmospheric nitrogen into combination. The union of nitrogen and oxygen under the influence of electric sparks, first discovered by Cavendish in 1785, and improved by Rayleigh in 1897 for preparing argon, was finally brought to fruition as a method for manufacturing nitrates by Birkeland and Eyde about 1903, who used the hydro-electric power available in Norway. As the union of the gases is a strongly endothermic reaction, they understood the necessity of causing combination to take place at the very high temperature of the arc, followed by rapid cooling of the product by imposing powerful magnetic fields at right angles, so as to cause the flame to be deflected and thus chilled as rapidly as possible to avoid undue decomposition of the nitric oxide. This compound is then caused to unite with more oxygen at lower temperatures, and the resulting nitrogen tetroxide absorbed by alkalis. A better method of nitrogen fixation, namely, by the direct union of the gas with hydrogen to form ammonia, was first made a success on a technical scale by Haber in 1910. As the reaction is exothermic, but takes place very slowly at low temperatures even in the presence of a catalyst, Haber was able to combine the opposing effects of heat by working at a temperature of about 500° C., and as combination takes place with diminution of volume, by employing a pressure of about 200 atmospheres. The effect of pressure in raising the yield of ammonia was further investigated by Claude in 1920, who succeeded in obtaining yields of about 40 per cent of ammonia by working the process at pressures of the order of 1000 atmospheres.

Although chemical equilibria in solution are determined primarily by the temperature and the effective concentration of the dissolved substances, it was pointed out by J. J. Thomson in 1888 that equilibria should be affected by capillary forces. He summarized his theoretical discussion of this subject in the statement that 'if the surface tension increases as the chemical action goes on, the capillarity will tend to stop the action, while if the surface tension diminishes as the action goes on, the capillarity will tend to increase the action'. Thomson was careful to point out that effects of this kind would predominate in thin films, but would not be apparent in the bulk of a liquid. Some forty years elapsed before a satisfactory experimental proof of this principle was obtained. In 1929, however, Freundlich directed attention to a phenomenon known as exchange adsorption which he considered was relevant to the problem. It was well known that charcoal when shaken with solutions of certain salts sometimes adsorbed one ion preferentially to the other. Freundlich also attached much importance to some experiments which Deutsch had carried out with indicators. It was found that solutions of indicators near their turning point changed their colours when shaken with an indifferent liquid such as benzene. This change of colour persists only as long as fine droplets of the organic liquid are emulsified in aqueous layer. The original colour is restored as soon as the two liquid layers have separated, but the process is strictly reversible. The change of colour produced by shaking corresponds to a numerical lowering of the pH value for an acid indicator like bromthymol blue; with a basic dye like brilliant green the effect of shaking is to produce a numerical raising of the pH value.

In 1884, van't Hoff published an important book entitled *Études de Dynamique Chimique* dealing with the course and influence of physical conditions on chemical change. In following up the experiments of earlier workers on chemical kinetics, a distinction was drawn between reactions of the first order, or unimolecular reactions, and reactions of the second order, or bimolecular reactions, and the distinction was regarded as fundamental. To the former class belong not only those reactions in which a single molecular species undergoes transformation, but the more numerous type in which one of the reactants is present in such enormous excess that its active mass is practically constant. Reactions of this latter type are now more suitably designated as pseudo-unimolecular reactions. Much attention was given to the problem of reconciling the order of a reaction as determined by experiments on the reaction velocity with the number of molecules which participate according to the chemical equation. The general result of considerations of this kind was in the direction of simplification, as many reactions of apparently complex character were found, when investigated by the methods of chemical kinetics, to be of the unimolecular or bimolecular type. A striking example was that of the reduction of potassium chlorate by ferrous sulphate in dilute sulphuric acid solution investigated by Hood (1878–85) and considered by him to be bimolecular. Van't Hoff recognized the possibility of reactions of higher order than the second, and in the second edition of his *Études*, an English translation of which was published in 1896, he quoted the results obtained by Noyes on the reduction of ferric chloride by stannous chloride as an example of a termolecular reaction. When the reaction takes place in the presence of hydrochloric acid the reaction is undoubtedly bimolecular, but when it is caused to take place in solutions as free from acid as possible it becomes termolecular. No wholly satisfactory explanation of this curious difference has been forthcoming. It may be noted that this result was obtained by what van't Hoff termed the variable volume method. At the outset he pointed out a fundamental distinction between first and second order reactions as regards the time required for a definite fractional transformation of the substances participating in a chemical change. The time required for half of a given amount of a substance to undergo transformation should be independent of the initial concentration if the reaction is of the first order, whereas if it is of the second order the time required for half-transformation is inversely proportional to the initial concentration. More generally the time required for half-transformation for an n-molecular reaction is inversely proportional to the $(n-1)$th power of the initial concentration. This important principle has been verified in innumerable cases, and indeed it is the best method for determining the order of a reaction. It will be seen in what follows that the more modern study of chemical dynamics has largely tended to regard the older and more rigid distinction between the different orders of reactions to be one of degree rather than fundamental, since it is frequently not difficult to vary the order of a reaction by appropriate variation of the experimental conditions.

A good illustration of the value of investigations of the kinetics of reactions is to be found in Bjerrum's experiments on the transformations of the various modifications of chromic chloride in solution. Two well-defined modifications of the hexahydrate of this salt had been known for many years, and some fundamental experiments relating to the reversible transformation of the violet salt, in which all the chlorine is in the ionic condition and is immediately precipitated by silver nitrate, into the dark green isomeride, in which only one-third of the chlorine is in the ionic condition, were carried out by Recoura in 1887. It was known that in the violet salt the six molecules of water of crystallization were very firmly held, whereas in the dark green salt two were readily removed, the other four being closely attached. The physico-chemical properties of these isomerides were fully studied by Werner and Gubser between 1901 and 1906, who were able to give satisfactory expression to their behaviour by formulating the violet salt as $[\mathrm{Cr}(\overset{+++}{\mathrm{H_2O}})_6]\,\overset{---}{\mathrm{Cl_3}}$ and the dark green salt as $[\mathrm{Cr}(\mathrm{H_2O})_4\mathrm{Cl_2}]^+\mathrm{Cl}^-,\,2\mathrm{H_2O}$. It will be noted that the co-ordination number of six is represented by the sum of the number of water molecules and chlorine atoms within the square brackets. In 1906–7, Bjerrum carried out an elaborate series of kinetic experiments on the transformation of the dark green salt into the violet salt in solution, the course of the reaction being followed by determinations of the conductivity and of the molecular extinction coefficient. He concluded that the reaction does not follow a simple first or second order rule, but consists of two consecutive first order reactions resulting in the formation of an intermediate hydrate. Bjerrum succeeded in preparing this intermediate compound—a light green salt in which two-thirds of the chlorine was found to be in the ionic condition, and one of the six molecules of water of crystallization was readily removed, and he accordingly assigned the formula $[Cr(\overset{+\,+}{H_2O})_5Cl]\overset{-\,-}{Cl_2}, H_2O$ to it. The mechanism of the transformation of the dark green salt into the violet isomeride takes place therefore in the following stages:

$$[Cr(\overset{+}{H_2O})_4Cl_2]\overset{-}{Cl} + H_2O = [Cr(\overset{+\,+}{H_2O})_5Cl]\overset{-\,-}{Cl_2}$$

and $$[Cr(\overset{+\,+}{H_2O})_5Cl]\overset{-\,-}{Cl_2} + H_2O = [Cr(\overset{+\,+\,+}{H_2O})_6]\overset{-\,-\,-}{Cl_3}.$$

These conclusions were confirmed by some experiments which Bjerrum carried out on the equilibria existing between the isomerides in solutions of different concentration and temperature, which greatly extended the earlier work of Recoura and others. It should be emphasized that the discovery of the light green isomeride was very directly due to the experiments on the kinetics of the transformations in solution.

Another interesting study on the kinetics of the transformation of a chromic salt in solution is to be found in the experiments of Freundlich and Pape, who investigated the hydrolysis of chloropentammine chromic chloride into aquopentammine chromic chloride in 1914. They followed the course of the reaction $[ClCr(\overset{+\,+}{NH_3})_5]\overset{-\,-}{Cl_2} + H_2O = [Cr(\overset{+\,+\,+}{NH_3})_5H_2O]\ \overset{-\,-\,-}{Cl_3}$ by measuring the flocculating power of the solution on the hydrosol of arsenious sulphide, the coagulative action of the bivalent purpureo salt being very much lower than that of the tervalent roseo salt. The results showed conclusively that the transformation of the purpureo into the roseo salt is a pseudo-unimolecular reaction.

In one notable instance, namely, in the transformation of ammonium cyanate into urea in aqueous solution, the reaction as studied by the methods of chemical kinetics is apparently more complex than might be expected on purely chemical grounds. The experiments of Walker and Hambly in 1895 have shown quite definitely that the rate of transformation is proportional to the square of the concentration of the salt, thus indicating a reaction of the second order. It is, however, possible that the reaction is concerned with the ions of the salt as distinct from the undissociated molecules, in which case the requirements of a bimolecular reaction are satisfied. This has been confirmed by other investigators, and it has been found that the reverse reaction, the transformation of urea into ammonium cyanate, is unimolecular.

Reference to the subject of autocatalysis has already been made in connexion with the experiments of Harcourt and Esson on the oxidation of oxalic acid by acidified potassium permanganate. Other examples are by no means uncommon. Thus the hydrolysis of an ester by water into the resulting alcohol and acid is catalyzed by the acid thus produced. A great deal of work has been done on esterification and hydrolysis, and it may be noted that the hydrolysis of esters is in general catalyzed both by acids and by alkalis, but the catalytic effect of alkalis is much greater than that of acids. In any case the processes of acid hydrolysis and alkaline saponification are somewhat different, the latter involving neutralization of the acid derived from the ester. The order of the reaction is in general different in the two cases, being first order for acid hydrolysis and second order for alkaline saponification.

The conversion of hydroxy-acids into lactones is catalyzed by the hydrogen ions of these acids, but there is clearly a difference between these two types of reaction, since in the hydrolysis of esters there is an accumulation of acid in the system, whereas in lactonization withdrawal of acid takes place. Much attention has been given to the general study of catalysis by acids, particularly with regard to the difficult problems of neutral salt action. It has become abundantly evident that catalysis by acids cannot be discussed in terms of the classical theory of ionization, since the effect of adding electrolytes has been found to exert an increase in the reaction velocity in some instances and a decrease in others. Among the numerous investigations of this subject, attention may be directed to the work of Brönsted about 1922–4, who introduced some conceptions which have tended to clarity in this difficult subject. Brönsted distinguished between what he termed the primary salt effect, which involved alteration of the activity coefficients of the solutes, and the secondary salt effect, which was concerned with the suppression of the ionization of weak acids, and he was able to support his theory by a considerable amount of experimental evidence.

A new chapter on the kinetics of reactions in which acids are in some way concerned may be said to have been begun with Lapworth's experiments on the bromination of acetone in 1904. This reaction proceeds as follows:

$$CH_3.CO.CH_3 + Br_2 = CH_3.CO.CH_2Br + H^+ + Br^-,$$

and its rate is directly proportional to the concentration of the acetone, but is independent of the concentration of the bromine. Further, the rate of this reaction is increased by the addition of either acids or bases, and precisely similar considerations are applicable to the iodination of acetone, a reaction which has received much attention at the hands of Dawson and his collaborators from 1909 onwards. It has been shown that the reaction, the velocity of which is actually measured, is the enolization of acetone, a relatively slow reaction, followed by a rapid reaction of the latter with the halogen, thus:

$$CH_3.CO.CH_3 \rightarrow (slow) \quad CH_2 : C(OH).CH_3 \rightarrow (with\ Br_2,\ rapid)$$
$$CH_3.CO.CH_2Br + H^+ + Br^-.$$

This work has been extended in many different directions by later workers, but it should be noted that the foundations of the study of the mechanism of reactions such as enolization, and particularly of esterification and hydrolysis, were laid down by Lapworth and his collaborators as early as 1908–12. Among the numerous and difficult problems which were attacked, mention may be made of the mode of formation of water when an alcohol is esterified by either a carboxylic acid or a sulphonic acid, and of the manner of fission of an ester when it is hydrolyzed by water. It will be evident that in esterification water may be formed either from the hydroxyl group of the alcohol and the hydrogen atom of the carboxyl or sulphonic group of the acid, or, alternatively, from the hydroxyl constituent of the acyl group of the acid and the hydrogen atom of the hydroxyl group of the alcohol. It has now been demonstrated by numerous experiments such as those of Ingold and his collaborators in 1939, who studied the hydrolysis of optically active esters containing an asymmetric carbon atom linking the alkyl group to oxygen, and those of Roberts and Urey in 1938, who investigated the esterification of benzoic acid containing excess of the heavy isotope of oxygen, that there is a difference between the processes of esterification and hydrolysis. In the esterification of a carboxylic acid the hydroxyl part of the carboxyl group unites with the hydrogen of the alcohol to form water and simultaneously oxygen from the alcohol is transferred to the ester, whereas in the hydrolysis of an ester the oxyalkyl group from the ester reacts with the hydrogen of the water to form the alcohol.

The numerous investigations involving the catalytic action of acids and bases have resulted in a broadening of view regarding the nature of these classes of compounds. An exaggerated emphasis was undoubtedly placed by the adherents of the classical theory of electrolytic dissociation upon regarding the properties of acids and of bases as exclusively concerned with hydrogen and with hydroxyl ions respectively, but it is, nevertheless, only right to point out that the study of these classes of substances in terms of the ionic theory enabled many phenomena to be understood much more clearly than could have been possible otherwise. This has been particularly noteworthy in certain departments of analytical chemistry. Thus such an expression as relative acidity employed by earlier generations of chemists acquired a more definite and quantitative meaning when stated in terms of hydrogen-ion concentration. This subject has been developed from several sides, and particularly in connexion with the study of indicators. Since about 1904, as the result of the work of Salm, Thiel, and others, it has been recognized that the sensitiveness of indicators can be expressed numerically in terms of the hydrogen-ion concentration corresponding to their change of colour. A few years later the practice of stating hydrogen-ion concentration in terms of Sörensen's familiar pH scale, in which the logarithmic value of the actual concentration with the negative sign neglected is used, came into general practice. Gradually opinion regarding the colour changes which are exhibited by indicators became more clearly understood. As long ago as 1894, Ostwald in his work *Die wissenschaftlichen Grundlagen der analytische Chemie* had ventured the opinion that the colour changes could be explained solely in terms of electrolytic equilibria. Others preferred to regard the colour changes as due to changes in constitution taking place on passing from the acid to the alkaline condition and vice versa. It is largely due to the investigations of Hantzsch on pseudo acids and pseudo bases since 1906 that it has become generally recognized that the elementary theory of indicators due to Ostwald is incomplete. In particular, it is recognized that a striking change of colour necessarily involves a change in the constitution of the ions. Indicators are thus to be regarded not as acids or bases which ionize *directly*, but as pseudo acids or pseudo bases, that is, as electrically neutral substances which are capable of undergoing tautomeric changes resulting in the formation of acids or bases, and these latter substances then ionize.

In 1923 an extended conception was given to acids and bases by definitions of these classes of compounds given by Lowry and independently by Brönsted. Acids were defined as substances which can lose hydrogen ions and bases as substances which can acquire hydrogen ions. In formulating this definition, Brönsted was careful to refer to the unhydrated hydrogen ion, the proton, and it will be clear that the terms are thus made to include substances which would not necessarily be considered as acids or bases according to the classical theory of ionization. An essential feature of the Brönsted-Lowry terminology relates to what they termed *conjugate* or *corresponding* acids or bases, and this does not depend upon the sign of the charge on an ion. Thus the ammonium ion NH_4^+ would be considered as an acid corresponding to the base ammonia NH_3, and the bisulphate ion HSO_4^- as an acid corresponding to the sulphate ion SO_4^{--}. This modern conception has been helpful in understanding the properties of acids and bases, as investigated by their catalytic effects on reactions, and by their action upon indicators in solvents other than water. Thus it may be noted that experiments on the relative strengths of acids in solvents such as ethyl alcohol have shown that the order of the acids is by no means identical with that observed when water is the solvent. It is also to be remembered that this extended conception of the nature of acids and bases appeared in the same year as the publication of the Debye-Hückel theory of complete ionization, in which the conceptions of the several deviation coefficients, viz. the activity, osmotic, and conductivity coefficients, were embraced in a comprehensive system, and thus enabled the difficult problems of chemical kinetics to be studied with more prospect of success.

It has already been remarked that the earlier investigators in the field of chemical kinetics attached an exaggerated importance to the order of a reaction as a measure of the number of molecules which participate in the reaction. The theory of the activation of molecules, originally stated in a somewhat indefinite manner by Arrhenius in 1889, has been developed into a theory of reaction velocity based upon the number of molecules colliding per unit of time. Three stages in the development of the collision theory have been recorded by Moelwyn-Hughes. First, there is the interpretation of the Arrhenius conception of the activation of molecules by correlating the velocity constant for a number of reactions with the absolute temperature. The Arrhenius equation can be rewritten to mean that the number of molecules undergoing chemical change per second is equal to a constant $\times e^{-E/RT}$. In this equation E is termed the critical increment of energy, and is either equal to, or in some manner approximates to, the energy of activation. The second step was taken in 1918 by Lewis who calculated velocity constants in absolute measure for reactions in gaseous systems. Thus taking Bodenstein's results for the rates of formation and of dissociation of hydrogen iodide, he was able to show that the number of molecules reacting per second is equal to the number of molecules colliding per second $\times e^{-E/RT}$. The third phase in the development of the subject centred round the difficult problem of unimolecular reactions. On *a priori* grounds it was by no means easy to accept a kinetic collision theory of the mechanism, since it would appear necessary for one molecule to collide with another in order to react with it, and thus it was not apparent how activation of the molecules in a unimolecular reaction could take place.

An alternative way of discussing this subject was, however, to be found in a general theory of chemical change, known as the radiation theory, and associated with the names of Lewis, Trautz, and especially of Perrin. This theory received much attention, particularly between the years 1918 and 1922, and it was found possible to discuss unimolecular reactions on the basis of the molecules becoming activated as the result of the absorption of radiant energy, and without any assumption of molecular collisions. This was particularly stressed by Perrin, who pointed out that since unimolecular reactions are characterized by the time required for a definite fractional change of the substance which undergoes transformation to be independent of the initial concentration, the probability that any molecule will decompose does not depend upon the impacts it receives. The radiation hypothesis was elaborated in various ways in order to discuss problems such as the effect of temperature on velocity of reaction. Perrin considered that the high temperature coefficient of the velocity of chemical reactions could be discussed on the basis of increase of radiation with rise of temperature as satisfactorily as on the assumption of increase in the kinetic energy of the molecules. Until the year 1922 the radiation theory of chemical change had enjoyed a very fair amount of support, but gradually a considerable amount of both theoretical and experimental evidence accumulated which conflicted with its requirements and thus showed it to be untenable. A way out of the difficulty regarding unimolecular reactions was, however, pointed out in that year by Lindemann, who suggested that molecules which decompose in a unimolecular manner could nevertheless be activated by collisions provided that the activated molecules persisted unchanged for a sufficient length of time, while some of them decomposed and others reverted to their normal condition by loss of energy. In this connexion it must be emphasized that the rate of activation and deactivation is very rapid in comparison with the rate of transformation of the activated molecules into the products of the reaction. A necessary consequence of this reasoning on kinetic grounds is that reactions involving gaseous decompositions taking place as reactions of the first order should approximate to reactions of the second order at sufficiently low pressures. Direct experimental confirmation of this conclusion was provided by Hinshelwood and his collaborators in 1926 and 1927 as a result of experiments on the thermal decomposition of various organic vapours, such as propionic aldehyde, acetone, and ether, and further experimental support from other investigators has shown the general accuracy of this theory of unimolecular reactions. The essential distinction between unimolecular and bimolecular reactions was stated by Hinshelwood in 1932 to consist fundamentally in the time elapsing between activation and reaction, and therefore in the importance of deactivation of active molecules by subsequent collision before chemical transformation has taken place.

Photochemical reactions have been the subject of very numerous researches, and no better illustration of the value of extended studies on a supposedly simple reaction is to be found than on the union of hydrogen and chlorine under the influence of light. This reaction was studied by Cruickshank, Dalton and others at the beginning of the nineteenth century. In 1843, Draper observed that chlorine which had been exposed to light appeared to become more chemically reactive towards hydrogen than it would be otherwise, and he ventured the opinion that the effect of exposure of the gas to light resulted in the formation of an allotropic modification of chlorine. This subject was investigated in detail by Bunsen and Roscoe in 1857, who were unable to confirm Draper's observations regarding the difference in reactivity shown by chlorine according as to whether it had or had not been previously exposed to light. It was, however, established by Draper and particularly by Bunsen and Roscoe that when a mixture of the two gases is exposed to light combination takes place gradually at first, but the rate of combination increases very considerably after a short time. This phenomenon was termed photochemical induction by Bunsen and Roscoe. An important observation regarding the effect of moisture on this reaction was made by Pringsheim in 1887. He showed that if the mixed gases are dried they are considerably less sensitive to the action of light than when they are moist, and concluded that combination between the gases does not take place directly, but through the formation of an intermediate compound in which water takes part. The existence of a period of induction could certainly be explained on the basis of the formation of a compound of this kind, possibly of chlorine monoxide produced by the action of the halogen on the water vapour. In 1902, Mellor showed that the phenomena of a period of induction could not be explained on the basis of the formation of chlorine monoxide, or of hypochlorous acid, as had been suggested by Becquerel and Frémy in 1879, since the addition of these substances to the mixed gases did not shorten the period of induction. Investigations carried out by Bevan in 1903, by Mellor, and by Burgess and Chapman between 1902 and 1906, have done much to clarify the nature of the period of induction. Bevan was of opinion that an intermediate addition compound of water and chlorine containing quadrivalent oxygen was first formed, and that this compound then reacted with hydrogen to produce hydrogen chloride and water. Burgess and Chapman were able to show that the existence of a period of induction was not a fundamental characteristic of the reaction, but was to be referred to secondary causes, and, in particular, that it was due to the presence of traces of impurities, such as ammonia, in the mixed gases. The presence of minute quantities of ammonia would result in the formation of nitrogen chloride which has a strongly retarding effect upon the reaction. It is interesting to note that van't Hoff in the *Études*, which was published many years previously, ventured the opinion that the existence of chemical induction or initial acceleration of chemical reactions is inconsistent with the fundamental principles of chemical dynamics, and must be referred to disturbing actions of some sort, and in more than one instance he was able to bring forward some experimental evidence in favour of the correctness of such a view. Subsequent investigators have confirmed the essential soundness of this conclusion.

Studies on the correlation of the amount of chemical change with the intensity of the light absorbed have led to results of great importance. Although there is much of interest to be found in the older experimental work since the time of Bunsen, it cannot be said that much progress was made as regards theoretical interpretation until the introduction of the quantum theory into chemistry. This theory, first enunciated by Planck in 1900, regards energy as discontinuous like matter or electricity and as radiated in definite units known as quanta. It originated as a result of certain difficulties which were encountered in the study of the thermodynamics of radiation.

Stefan in 1879 had pointed out that the rate of radiation is proportional to the fourth power of the absolute temperature, and in 1884 Boltzmann, adopting an idea due to Bartoli and using Maxwell's electromagnetic theory of radiation, deduced Stefan's law by the method of the Carnot cycle. On the experimental side the general accuracy of Stefan's law has been verified by various investigators, particularly by Lummer and Pringsheim since 1897, but there were certain difficulties regarding the distribution of energy from different parts of the spectrum which could not be overcome on the basis of classical thermodynamics, but which were resolved by the new ideas introduced by Planck. The quantum theory has been extremely fruitful in several widely different directions in physical chemistry, but for the present purpose the discussion must be restricted to a very brief application to photochemical reactions.

The fundamental principle of the quantum theory may be expressed by the equation $E = h\nu$, where E is the energy, ν the frequency of the radiation, and h is Planck's constant having a value of $6 \cdot 6 \times 10^{-27}$ erg second. A light quantum is frequently termed a photon. In any photochemical reaction at least one photon must be absorbed for every molecule which becomes activated, but it is nevertheless possible for molecules to become activated in other ways as well. Thus in the reaction between hydrogen and chlorine, Einstein's law of photochemical equivalence, according to which one molecule is transformed for every quantum of light which is absorbed, is not even approximately followed. On the contrary, Bodenstein in 1913 observed that one photon brings about the union of many thousands, possibly of millions of molecules, and he was thus led to suggest a theory of the reaction based upon what is now termed a chain mechanism. In 1918 greater precision was given to this idea by Nernst, who considered that the primary reaction consists in the photochemical dissociation of the chlorine molecules into atoms, followed by secondary reactions such as are represented by the equations $Cl + H_2 = HCl + H$, and $H + Cl_2 = HCl + Cl$, these latter reactions taking place without the absorption of fresh photons. But the chlorine atoms produced in one of the secondary reactions can react with fresh hydrogen molecules, and thus set up a chain and thereby lead to the formation of an enormous number of molecules of hydrogen chloride for the original quantum of light which was absorbed. A chain of this kind can be broken by the direct formation of molecules of hydrogen, chlorine, and of hydrogen chloride from the free atoms. A vast amount of work has been done on this reaction since Nernst suggested a chain mechanism for it, and the reaction is evidently more complicated than was at first supposed, but the fundamental accuracy of Nernst's theory has been verified, particularly as the results of the experiments of Norrish and Ritchie in 1933, and of Chapman and his collaborators over a number of years.

The theory of reaction chains has been extended to numerous other types of reactions, some of which take place in the gaseous phase and others take place in solution. Before proceeding to discuss some aspects of this subject it should be noted that although the theory of chain reactions was originally introduced to explain the enormous number of molecules of hydrogen and chlorine which are caused to react as the result of the absorption of a single quantum of light, the idea of a chain mechanism was also introduced independently in 1923 by Christiansen and Kramers to overcome certain difficulties connected with the kinetic theory of unimolecular reactions. It was remarked by Semenoff that this interesting idea failed to attract much attention because the new ideas were applied to old material, namely, the decomposition of nitrogen pentoxide. In the following year, however, Christiansen applied the principle to develop a theory of negative catalysis based upon the idea of the destruction of the chains. It will be seen that many curious phenomena which have long been known in connexion with various chemical reactions, particularly with the processes of oxidation and reduction, are capable of being discussed in terms of a theory of this kind.

Investigations by Schönbein and others during the nineteenth century had shown that in many instances of oxidation at ordinary temperatures the process is by no means a simple and straightforward union of the oxidizable substance with oxygen, but that some of the oxygen becomes converted into ozone or into hydrogen peroxide during the process. This interesting subject, known as autoxidation, embraces subjects such as the formation of ozone during the oxidation of phosphorus and the production of hydrogen peroxide when certain metals become oxidized in the presence of moisture. In many instances Schönbein was able to observe that the oxygen apparently divides itself into two equal parts, one part being concerned with the oxidation of the substance, and the other part being activated by conversion into ozone or hydrogen peroxide. This subject was afterwards studied by a number of chemists, particularly by van't Hoff, Jorissen, Engler, Traube, Bach, and others, and the general accuracy of Schönbein's observations was confirmed and extended. Thus in 1900, Engler demonstrated the formation of hydrogen peroxide when a hydrogen flame was quenched by impinging upon ice, but an experiment of this kind does not give much indication as to whether the hydrogen peroxide is formed directly, or is a secondary product of the reaction. In attempting to formulate a theory of autoxidation Schönbein favoured some views expressed by Brodie between 1850 and 1864, according to which the molecules of ordinary oxygen in becoming activated give rise to what were termed ozone and antozone, these latter substances being endowed with electric charges of opposite sign. A theory of this sort appears to have arisen, in part at any rate, as a result of studies on reactions such as the decomposition of hydrogen peroxide by acidified potassium permanganate or by silver oxide, or by alkaline potassium ferricyanide, in which two powerful oxidizing agents destroy each other with evolution of molecular oxygen. Derivatives of hydrogen peroxide were regarded as antozonides, and compounds such as silver oxide were regarded as ozonides. Although it was largely due to Brodie that derivatives of hydrogen peroxide such as diacetyl peroxide were first prepared and studied, and thus gave rise to more modern developments in connexion with the true per-acids, the Brodie-Schönbein theory did not survive for very long. Thus the supposed distinction between oxonides and antozonides was difficult to maintain, since barium peroxide was found to yield either hydrogen peroxide or ozone according as it was treated with dilute or with concentrated sulphuric acid.

A different type of theory was advanced independently by Engler, Bach, Manchot and others between 1897 and 1902.

According to these chemists the first effect of the action of an oxidizing agent or of molecular oxygen upon a substance is to produce a higher oxide, what Manchot has termed a primary oxide, which subsequently breaks down to yield the normal oxidation products of the reaction, and this may involve the production of hydrogen peroxide. It is not difficult with the aid of a theory of this kind to understand an interesting result obtained by Jorissen in 1897. He found that whereas dilute solutions of sodium arsenite are stable to atmospheric oxygen, solutions of sodium sulphite are rapidly oxidized under these conditions, but if oxygen is allowed to act upon a solution of an arsenite and a sulphite, *both* salts are simultaneously oxidized. Other examples of interest in volumetric analysis are well known. Thus it has long been known that chlorine is not evolved from very dilute solutions of hydrochloric acid and potassium permanganate at the ordinary temperature, but if a ferrous salt is present it becomes oxidized to the ferric condition, and simultaneously some of the hydrochloric acid becomes oxidized to chlorine. Manchot has given a fairly satisfactory explanation of these phenomena in terms of his theory of the formation of a primary higher oxide of iron.

There is a considerable amount of experimental evidence which goes to show that in different types of reactions involving oxidation, the first effect of the attack of the oxidizing agent is to produce some higher oxide. Thus when the alkali metals are burnt, the product is invariably a peroxide, never the normal oxide. A most interesting series of experiments was carried out by Job since 1899 on the action of atmospheric oxygen on solutions of potassium carbonate containing a cerous salt. He found that if a solution of a cerous salt is added to a concentrated solution of potassium carbonate with free access to the air, the liquid assumes a dark red colour due to the formation of a perceric carbonate. Job obtained a crystalline compound to which he assigned the formula $Ce_2(CO_3)_3O_3, 4K_2CO_3, 12H_2O$ in this way. Solutions of this perceric potassium carbonate were found to be highly efficient carriers of oxygen to substances which are not easily oxidized directly. Thus arsenites are readily oxidized to arsenates, and glucose and other reducing sugars are oxidized under these conditions. According to Engler the formation of the perceric compound does not take place directly, but through the intermediate formation of hydrogen peroxide. The catalytic action of cerous salts in effecting oxidations is not unlike that of ferrous salts. An interesting series of such reactions effected by hydrogen peroxide in the presence of ferrous salts was carried out by Fenton between 1894 and 1902. Thus, he showed that tartaric acid is oxidized to dihydroxymaleic acid, and in more recent times the catalytic action of iron has been studied in detail by Wieland and others. In 1924, Goard and Rideal investigated the oxidation potentials of solutions of sodium arsenite under the influence of a cerous salt dissolved in concentrated solutions of potassium carbonate. They also examined the oxidation of hydriodic acid by hydrogen peroxide in the presence of ferrous salts—a reaction known to Schönbein—in the same way. In the course of this work they concluded that the formula of the primary oxide of iron which is formed in reactions of this kind is Fe_2O_5, the same formula which was suggested by Manchot in 1902.

In connexion with the subject of autoxidation and of induced reactions generally, attention may be directed to the terminology which has been introduced since 1903, chiefly in consequence of the suggestions of Luther and Schilow. The oxidizing agent, e.g. atmospheric oxygen, hydrogen peroxide, potassium permanganate, or chromic acid, is termed the *actor*, the compound which is readily oxidized under almost any conditions, e.g. sodium sulphite, ferrous salts, etc., is termed the *inductor*, and lastly the compound which is oxidized when both actor and inductor are present together, but is not normally oxidized by the actor alone, e.g. sodium arsenite, tartaric acid or other hydroxy compounds, is termed the *acceptor*.

As is well known oxidation is concerned not only with the addition of oxygen but also with the removal of hydrogen from substances. A very remarkable example of this is to be observed in the action of phenylhydrazine acetate upon aqueous solutions of aldose or ketose sugars. The hydrazone which is first formed is transformed into an osazone, the process involving oxidation of the adjacent alcohol group to a carbonyl group by the phenylhydrazine withdrawing two atoms of hydrogen resulting in the formation of aniline and ammonia, and thus acting as an oxidizing agent. According to Wieland the latter aspect of the subject is of greater importance than the former. Thus in 1912 he showed that carbon monoxide and water in the presence of palladium black react to produce carbon dioxide and hydrogen, the oxidation of the carbon monoxide to the dioxide being effected by the removal of the hydrogen by the palladium. Wieland also regarded the numerous dehydrogenations studied by Sabatier and Senderens from this point of view. He found that formic acid is produced in the moist oxidation of carbon monoxide, presumably as an intermediate compound which subsequently breaks down to carbon dioxide and hydrogen. Wieland regarded oxidations which take place by ordinary oxygen as due to the unsaturated character of the molecule of the gas, thus enabling it to react by addition, and this may result in the formation of hydrogen peroxide if a compound containing hydrogen is attacked. This subject was studied by von Wartenberg and Sieg in 1920, using a hot-cold tube method, and they found that in the union of hydrogen and oxygen, hydrogen peroxide is *directly* formed at temperatures between 600 and 1000° C., and this compound then decomposes into water and oxygen. The results which they obtained in the oxidation of moist carbon monoxide were in essential agreement with those of Wieland, namely, that the primary product is formic acid formed by the direct union of the gas with water; the formic acid is then oxidized to carbon dioxide and hydrogen, the hydrogen forms hydrogen peroxide directly, and lastly, the hydrogen peroxide decomposes into water and oxygen, the whole process being of the nature of a chain reaction.

Willstätter and Waldschmidt-Leitz in 1921 investigated the hydrogenation of some organic compounds in presence of metals, and concluded that the presence of oxygen is necessary to account for some of the effects which were produced. They found that platinum and palladium when completely freed from every trace of oxygen were incapable of effecting the hydrogenation of readily reactive olefinic compounds, which are readily reduced to saturated compounds under ordinary conditions. Thus catalytic hydrogenation was regarded by these authors in a wholly different way from any theory of the formation of unstable metallic hydrides as viewed by Sabatier and Senderens and by Wieland.

The oxidation of sodium sulphite in solution has been a fruitful subject of investigation in connexion with theories of chain reactions and of catalysis generally. Titoff in 1903 observed that the reaction is markedly catalyzed by minute traces of copper salts, and also confirmed the observations which others had previously made that the reaction is strongly retarded by a number of substances, such as certain alcohols and other organic compounds. Bäckström in 1927 showed that the subject can be discussed in terms of Christiansen's theory of negative catalysis based upon the conception of destruction of chains. As the results of kinetic experiments, Bäckström was able to show that the retarding effects of mannitol, methyl alcohol, ethyl alcohol, and benzyl alcohol were in the ratio of 1 : 0·5 : 1·3 : 60. Alyea and Bäckström in 1929 showed that benzaldehyde is formed when benzyl alcohol is used as the inhibitor in the autoxidation of sodium sulphite, and similarly acetone is formed when *iso*propyl alcohol is used. The destruction of the chain is thus effected by the oxidation of the substance which acts as the negative catalyst. The correctness of the theory of a chain mechanism is strongly supported by photochemical investigations, since one quantum of light effects the oxidation of many thousands of molecules of sodium sulphite.

It may be noted that most reactions which involve autoxidation can be discussed in terms of the theory of the formation of a primary oxide or, alternatively, in terms of the more modern ideas of the formation and destruction of chains, and indeed the two points of view are by no means antagonistic, but rather complementary to each other. Thus many familiar phenomena, such as the inhibition of the glowing of phosphorus in air by the presence of very minute quantities of certain organic compounds, which was familiar to Schönbein, and the relative stability of rongalite, the formaldehyde derivative of sodium sulphoxylate, as compared with the very powerfully reducing sodium hydrosulphite, are capable of being understood from these points of view. Coming to more modern studies on oxidation, particular importance should be attached to a long series of investigations by Moureu and Dufraisse since 1919. These experiments originated in connexion with observations on the polymerization of acraldehyde to disacryl. They found that the autoxidation of acrolein, as well as its polymerization to disacryl, was strongly retarded by readily oxidizable organic compounds such as hydroquinone and pyrogallol. Similar considerations are relevant to the inhibition of the oxidation of other aldehydic substances. Thus in 1924, Moureu and Dufraisse showed that the autoxidation of benzaldehyde could be completely stopped by the addition of one-thousandth of the quantity of hydroquinone. They pointed out that the kinds of substances which inhibit oxidation are themselves readily oxidizable.

Autoxidation phenomena and the formation of peroxides generally seem to exert influences in wholly unexpected directions. As long ago as the year 1870, Markownikoff formulated certain rules relating to the addition of the hydrogen halides to unsaturated compounds. Thus he stated that when hydrogen bromide is added to an olefine, the bromine atom always goes to the carbon atom having the fewest number of hydrogen atoms attached to it. The addition of hydrogen bromine to propylene thus results in the formation of *iso*propyl bromide, not normal propyl bromide. Since 1933 it has been shown by Kharasch and his collaborators that the nature of the products which are obtained in reactions of this kind is dependent on whether peroxides are present or not. In general, it appears to have been established that when peroxides are rigidly excluded, Markownikoff's rule is strictly obeyed, but if certain peroxides are present the addition of hydrogen bromide to propylene results in the formation of normal propyl bromide, the peroxide having therefore a directive catalytic action. It should be noted that these effects are not observed with hydrogen iodide, presumably because of the powerfully reducing properties of this compound which would result in the removal of any peroxides.

In bringing this brief sketch of some of the more distinctive features of chemical change to a conclusion, it will have been observed how much progress has been made as the result of the investigation of old and familiar reactions by modern methods. Experimental refinements have been accompanied with advances in theoretical ideas. Classical thermodynamics and the kinetic theory of gases have enabled fundamental principles to be firmly established, but it may be added that the more recent developments have arisen much more in consequence of the application of kinetic theory than of thermodynamics. The limitations of classical thermodynamics have been amplified by the introduction of the quantum theory—just one more example of the tendency to atomize our processes of reasoning. By these various means the theories of the activation of molecules and of chain reactions—the corner stones of modern chemical dynamics—have been placed upon a secure foundation.

REFERENCES

J. H. VAN'T HOFF. *Studies in Chemical Dynamics*. Translated by T. Ewan. Amsterdam, 1896.

J. J. THOMSON. *Applications of Dynamics to Physics and Chemistry*. London, 1888.

J. W. MELLOR. *Chemical Statics and Dynamics, including the Theories of Chemical Change, Catalysis, and Explosions*. London, 1904.

W. NERNST. *Experimental and Theoretical Applications of Thermodynamics to Chemistry*. London, 1907.

E. A. MOELWYN-HUGHES. *The Kinetics of Reactions in Solution*. Oxford, 1933.

C. N. HINSHELWOOD. *The Kinetics of Chemical Change*. Oxford, 1940.

LOUIS S. KASSEL. *The Kinetics of Homogeneous Gas Reactions*. New York, 1932.

K. C. BAILEY. *The Retardation of Chemical Reactions*. London, 1937.

N. SEMENOFF. *Chemical Kinetics and Chain Reactions*. Oxford, 1935.

H. B. WATSON. *Modern Theories of Organic Chemistry*. Second edition. Oxford, 1941.

E. J. BOWEN. *The Chemical Aspects of Light*. Oxford, 1942.

LOUIS P. HAMMETT. *Physical Organic Chemistry*. New York, 1940.

H. D. O'SULLIVAN. *The Life and Work of C. O'Sullivan*, F.R.S. Guernsey, 1934.

R. WILLSTÄTTER. Problems and Methods in Enzyme Research. Faraday Lecture. *J. Chem. Soc.* 1927, p. 1359.

H. FREUNDLICH. Surface Forces and Chemical Equilibrium. Second Liversidge Lecture. *J. Chem. Soc.* 1930, p. 113.

I. M. KOLTHOFF. *Acid-Base Indicators*. Translated by C. Rosenblum. New York, 1937.

C. MOUREU and C. DUFRAISSE. Catalysis and Autoxidation. *Chem. Rev.* 1927, III, 113.

INSTITUT INTERNATIONAL DE CHIMIE SOLVAY. Deuxième Conseil de Chimie: *Structure et Activité Chimique*. Paris, 1926. Cinquième Conseil de Chimie: *L'Oxygène et ses Reactions Chimiques et Biologiques*. Paris, 1935.

Annual Reports of the Chemical Society since 1904.

Chapter IX

A RETROSPECT

In viewing the vast and ever-extending growth of modern chemistry, and thus attempting to compare it with the position in the early years of the nineteenth century, it will soon become apparent that the contrast is by no means one of *magnitude* only. There are other though perhaps less immediately obvious differences—differences of emphasis and of relevancy which become more clearly defined when the subject is studied in its true perspective. Perhaps one of the most distinctive features of the development of chemistry from these earlier times has been its gradual transformation from a largely descriptive science into an 'exact' one. The emphasis has been very markedly in the direction of what might be termed *kinetic* chemistry. This process, in which much of the methodology of chemistry has approached more closely to that of physics, has been necessarily accompanied with a greatly increased importance of the theoretical aspects of the science. Indeed, the history of the relations between chemistry and physics from the early years of the nineteenth century to the present time is of peculiar interest. From early times the two sciences were largely developed together, and fundamental discoveries in both were frequently made by the same investigators. Thus men like Cavendish, Faraday, Hittorf, and Bunsen enriched both departments of study by their researches. Gradually, however, the two sciences drifted apart, and for this separation various explanations may be suggested. It would appear certain that the directions in which physics was being chiefly developed during the nineteenth century by men such as Clerk Maxwell, Kelvin, and Stokes were not such as to arouse much interest among chemists, who were then much concerned with developing the atomic theory and embarking upon the vast territories of organic chemistry. Towards the close of the nineteenth century, however, the two sciences came together again, and the years 1885–7, in which the foundations of the theory of dilute solutions were laid by the publication of van't Hoff's papers on osmotic pressure and Arrhenius's papers on electrolytic dissociation, may be said to mark the starting of physical chemistry as a special department of study. Since that time the two sciences have become ever more closely associated, and the study of physical chemistry has exerted a profound influence upon the development of both inorganic and organic chemistry.

Although the influence of physics upon the development of chemistry is disputed by no one, the history of the science affords instances in which progress has been delayed or impeded by the reluctance of many chemists to accept results obtained by physical methods. This assumption of an attitude of caution, amounting in some instances to one of actual opposition, to physical methods of investigation was characteristic of much of nineteenth-century chemistry. Thus the opposition of many chemists to Avogadro's rule was to some extent an antagonism to an essentially physical theorem—it took nearly fifty years to become a fundamental part of the framework of chemistry, particularly as the result of the activities of Gerhardt and especially of Cannizzaro. At the same time it cannot be denied that a number of researches of fundamental importance, which have left permanent marks on the science, were carried out during a period when chemical theory was in a most unsatisfactory way. Even before the general adoption of Cannizzaro's system of atomic weights, and modern formulae had come into general use, much progress had been made in electrochemistry, and in organic chemistry important beginnings had been made in a number of directions. Bunsen's researches on the cacodyl compounds belong to this period, which also includes the discovery of the pyridine bases by Anderson (1848–54). Progress in such numerous and varied directions may even suggest if an exaggerated value is not sometimes placed upon theoretical considerations. Erroneous theories have sometimes been of much value in the development of physical science. Thus the corpuscular theory of light, aptly described by Larmor as 'actually on wrong lines and not merely incomplete, was yet a useful hypothesis in its day,...that it served as a basis for great practical advances in astronomical and optical science', might be quoted as an example.

Interesting examples of the hesitation of chemists to welcome results obtained by physical methods are to be seen in the application of physical properties to problems of constitution. Thus the subject of molecular refractive power, first studied empirically by Gladstone and Dale in 1858 and afterwards based upon the electromagnetic theory of light by Lorentz and Lorenz about 1880, has been shown by Brühl and others to be a constitutive property, and therefore a valuable aid in attempting to solve problems of constitution. Although some of the results, particularly with conjugated unsaturated linkages, have been found difficult to interpret, the neglect with which the subject was treated was unjustified. The long and bitter controversy which took place between Hantzsch and Bamberger since 1894 regarding the constitution of aromatic diazo compounds is a good example of the value of physico-chemical methods of investigation. Bamberger endeavoured to establish the constitution of these highly reactive substances by the aid of purely chemical methods, whereas Hantzsch invoked information derived from the electrical conductivity of the salts. Thus Hantzsch was able to show that the diazonium salts are good electrolytes and not appreciably hydrolyzed, and also that the isomerism of the diazotates could be discussed on a stereochemical basis similar to that of the oximes. It is true that contributions of value were also made by Bamberger, but the views regarding the constitution of diazo compounds generally which are held at the present time are fundamentally those put forward by Hantzsch. One more example might be quoted. After the discovery of argon by Rayleigh and Ramsay in 1894 and of helium and the other inert gases a few years later, the evidence of the monatomic character of these gases rested on the ratio of the two specific heats having a value of 1·67. Although a value of this kind was demanded by the kinetic theory for monatomic gases, and had actually been obtained for mercury vapour long before, chemists were by no means satisfied with evidence of this kind to formulate the molecules of helium and argon. As no gas known to be diatomic has ever been found with a ratio for the specific heats higher than 1·4, everyone was gradually convinced of the value of the conclusions which were drawn in this way.

In the twentieth century opposition to chemical results derived by physical methods was much less pronounced. It would appear that the value of physical methods was not only more generally recognized, but the kind of physics which was receiving attention from investigators at that time was such as to be of much greater interest to chemists than the older physics. The study of the conduction of electricity through rarefied gases and of radioactive phenomena were very much in the ascendant, and resulted in the recognition of electrons as universal constituents of matter, and that radioactive change could only be explained in a consistent and satisfactory manner as due to atomic disintegration. Far-reaching results of this kind undoubtedly had the effect of bringing the two sciences more closely together.

The influence of physics upon the study of chemistry can be to some limited extent judged by the style of the standard text-books in use at various times. In the nineteenth century the works on inorganic chemistry were almost wholly of a descriptive character and, it might be added, with very little in the way of any theoretical foundation. A noteworthy exception, however, was the celebrated *Principles of Chemistry* by Mendeleeff, in which the subject was developed around the periodic law. General and physical chemistry was largely treated as a separate department of study, and its chief principles were to be found in Lothar Meyer's *Modern Theories of Chemistry*. At a later date, however, the various writings of Ostwald, such as his *Principles of Inorganic Chemistry*, an English translation of which appeared in 1902, exerted a profound influence upon the study of inorganic chemistry. In these works general principles, such as the theory of electrolytic dissociation, the phase rule, and the theorem of Le Chatelier, were emphasized, and their application discussed with constant reference to subjects which in other works were treated in a purely descriptive manner. Much of the value of Ostwald's writings was, however, diminished as the result of his antagonism to the atomic theory which he regarded as of too hypothetical a nature to be a suitable foundation for chemical science. Ostwald preferred to use energetics as a basis for the development of the subject, as he attempted to show in his Faraday Lecture in 1904. The attempt was foredoomed to failure, but it is not without significance that four years later when engaged in revising one of his earlier works, Ostwald recognized the work of J. J. Thomson and others on electrons, of Rutherford and Soddy's studies on radioactivity, and of Perrin's experiments on the Brownian movement, and was thus at last compelled to admit the physical reality of atoms and molecules. It should be noted that the standard works of Nernst and of van't Hoff on physical chemistry, which were written at about the same time as some of Ostwald's treatises, were altogether free from this prejudice against the atomic theory. The title-page of Nernst's celebrated *Theoretical Chemistry*, the first English translation of which appeared in 1895, emphasizes its principles in words which require no comment, viz. *from the standpoint of Avogadro's rule and thermodynamics*. The rapid progress of modern atomic science has been well reflected in more recent text-books on inorganic chemistry. In particular, the introduction of X-ray spectroscopy by Moseley in 1913, which resulted in the recognition of the atomic number of an element as distinct from its atomic weight as the more fundamental unit and thus illuminated the periodic law, and the developments which have followed the experimental study of isotopes and of the X-ray analysis of crystals have received ample attention in modern works.

The necessity for chemists to become endowed with a knowledge of certain branches of physics has become increasingly evident, particularly in the last few decades. Chemists with a relatively limited acquaintance with physics were by no means debarred from entering the classical school of physical chemistry, but the position is very different at the present time. Up to the beginning of the present century the needs of chemists were amply satisfied with classical thermodynamics and the kinetic theory of gases. Thus all the fundamental principles of chemical statics and kinetics and the classical theory of electrolytic dissociation were capable of discussion in terms of these physical principles. The introduction of Planck's quantum theory in 1900 in order to clarify certain anomalies connected with the thermodynamics of radiation was destined to lead to far-reaching results in physical chemistry. Thus Bohr in 1913 was able to effect certain improvements in Rutherford's atomic model by means of quantum mechanics, and the modern study of photo-chemical reactions would have made no progress without the conception of the absorption of radiant energy in definite quanta of light or photons as these have been termed. Another direction in which the quantum theory has been of much value is in connexion with specific heats. Most of the theory of the specific heats of gases, such as the dependence of the value of the ratio γ on the atomicity of any particular gas, was derived from the principles of the kinetic theory. The problems connected with the specific heats of solids are more difficult. It was not until after the formulation of the quantum theory that an understanding of Dulong and Petit's empirical law and of the variation of the value of the specific heats of solids with temperature received a satisfactory explanation. Much of the experimental work connected with this subject was carried out by Nernst and Lindemann about 1911.

Jeans has pointed out certain limitations of the kinetic theory of gases and in particular why the analogy between billiard balls and molecules fails as soon as questions of internal vibrations and the transfer of their energy to the outside are considered. The reason for this failure is because the motion of objects of the order of magnitude of billiard balls is governed by Newtonian dynamics, whereas the internal motions of molecules and the transfer of their energy to the external space are governed by an entirely different set of laws. The modern theory of quantum mechanics originally developed by Louis de Broglie since 1922 has been eagerly studied in subsequent years, and one of its most valuable results as regards chemistry is to be seen in the modern theory of resonance, associated more particularly with Pauling. As applied to benzene the theory of resonance regards the bonds between the carbon atoms to be all alike and intermediate in properties between single and double bonds, and thus provides a means of explaining the peculiarities of aromatic character.

In viewing the development of modern atomic science it should not be forgotten that chemistry as well as physics has made a substantial contribution. Soddy in 1935 pointed out that the modern conception of atomic number dates back to the displacement law of radioactive change associated with the names of Fajans, Russell, and himself in 1913. This is really a chemical law concerning the change of character of an element resulting from the expulsion of an α-particle or a β-particle from its atom, and was elaborated by Soddy in his scheme for incorporating the radio-elements into the periodic law shortly before Moseley's first publication on atomic numbers. Moseley's generalization with regard to atomic numbers was the same as that of Aston with regard to isotopes, namely, an extension of a conception which had originated in radio-chemistry to all the elements in the periodic table. It might be added that although there has perhaps been a tendency on the part of physicists to overlook the contributions which have been made by chemists, it is not difficult to see how such a state of affairs could arise. The reason is simply because chemists working in the field of atomic science require to be much more abundantly endowed with a knowledge of physics than have physicists to be equipped with a knowledge of chemistry.

The history of the relation of organic chemistry to biological science is in some respects not unlike that of physics to chemistry, namely close collaboration between the two departments of study in the early stages followed by a period of separation and later by the two sciences becoming ever closer together. The beginnings of organic chemistry were necessarily directed to the synthesis of vital products and thus to the overthrow of the theory of vitalism. As the subject developed, largely under the influence of Liebig and Wöhler, the necessity for a logical scheme for the discussion of the problems of constitution became of prime importance. In his Faraday Lecture in 1907, Emil Fischer remarked that three of the ablest of Liebig's pupils, namely, Hofmann, Wurtz, and Kekulé, abandoned the biological aspects of organic chemistry altogether and devoted themselves to researches which were destined to be of great theoretical importance. Indeed, the departure from biology was a necessary stage in the evolution of this branch of science while experimental methods and theories were being elaborated.

It has been remarked that although the synthesis of ethyl alcohol from ethylene, effected by Hennell, and that of urea by the intramolecular transformation of ammonium cyanate, accomplished by Wöhler, were discoveries of first-rate importance, neither of them could with strict logic be claimed as having *completely* overthrown the theory of vitalism. Neither of these syntheses was 'complete', because the cyanate used by Wöhler had been derived from alkali cyanides obtained from nitrogenous organic matter, and Hennell's olefiant gas had been obtained by the thermal decomposition of oil. This criticism is no longer valid since both ethylene and cyanides can be prepared from non-vital sources. From the biochemical standpoint it is by no means sufficient to have accomplished the preparation by laboratory methods of some particular vital product. It is true that many such products have been synthesized in the laboratory, but how many, or should it be said how few, have been prepared under conditions such as prevail in the organism of the animal or plant? Moreover, it is by no means certain whether many naturally occurring products have been formed in the organism by synthetical processes or have resulted from the breakdown of more complex substances. The ultimate aim of biochemistry, as Fischer pointed out, is to gain a complete insight into the unending series of changes which attend plant and animal metabolism.

Much progress has been made in recent years in the synthesis of natural products, and there can be no doubt that the biochemical aspects of organic chemistry are certain to become of ever-increasing importance. Much of the progress in work of this kind has resulted from improvements in experimental methods for dealing with small quantities of substances, and particularly to developments in microchemical methods of analysis. Nor should it be forgotten that some advances in physical chemistry have arisen in consequence of investigations in biological science. The beginnings of the study of osmotic pressure were made chiefly by plant physiologists such as Pfeffer and De Vries, and there is no doubt that their researches had a profound influence upon the development of van't Hoff's theory of dilute solutions. This theory and the classical and modern theories of ionization have had much to do with some of the developments in physiology and biochemistry, more especially in connexion with phenomena which are caused by the semipermeable membranes which occur in living cells. Another example which might be quoted is the widespread attention which has been given to the study of hydrogen-ion concentration in recent years. This subject has been very much developed as the results of Sörensen's studies on the rates of enzyme action and the stability of proteins since 1909. He found that these phenomena are considerably affected by the relative acidity of the solutions, and this gave rise to numerous electrochemical researches on the determination of hydrogen-ion concentration and investigations on the properties of indicators.

Such various interchanges between different departments of natural knowledge have been the means of progress in widely different directions. In this connexion the history of chemistry has provided ample evidence of the exaggerated importance which has far too often been attached to distinctions between pure science and applied science. Great advances in industrial chemistry have resulted from the direct application of physico-chemical principles to manufacturing operations, as may be seen in an industry such as the fixation of atmospheric nitrogen, where progress has been very much a consequence of studies on the kinetics and equilibria of gaseous reactions. The rare gas industry has been developed eventually from Rayleigh's work on the difference which he observed between the density of nitrogen derived from the atmosphere and from its compounds. Numerous other examples might be quoted, but it is interesting to turn to the other side of the picture and see that developments in pure physics and chemistry have sometimes arisen from research on technical problems. Carnot, the founder of thermodynamics, was led to his *Réflexions* as a result of his desire to improve steam and other heat engines, and some of the progress in the study of enzymes can be traced directly to problems which have arisen in the brewing industry, of which the research on invertase by O'Sullivan and Tompson in 1890 might be quoted as an example.

The history of chemistry has provided ample evidence of the unavoidable separation of the subject into various specialized departments. But this very process has sometimes thereby led to a temporary neglect of phenomena on the borderland of these departments. Eventually a new kind of specialization has arisen. Thus the studies on subjects such as the mechanism of organic reactions have become included in what has been termed physical organic chemistry. Another remarkable instance of a change of emphasis has come over analytical chemistry. This branch of the subject was at one time regarded as the handmaid of pure or applied chemistry, and a few years ago it was designated as applied physical chemistry—both descriptions might be regarded as equally inaccurate. Analytical chemistry has made vast progress in recent years, as may be seen in the rapidly extending use of organic reagents in inorganic analysis, the widespread interest in the study of indicators, and in the ever-increasing introduction of various physical—electrical, spectroscopic, and radioactive—methods into analytical processes.

The present position of chemistry was well described in the words with which Mills concluded his Presidential Address to the Chemical Society in 1944. 'Chemistry is more interesting now than it ever was. There is more logical connexion between the facts. But the amount the student has to learn is greater and the temptation to over-specialize much increased. Our deeper knowledge enables us now to see things in truer perspective, and the teacher needs more than ever to seek out the essentials and to find the simplest methods of presentation and so to help the student on the long road he has to travel to the stage where he can undertake independent research.'

INDEX OF NAMES

INDEX OF SUBJECTS